Salt-Water Purification

Salt-Water Purification

SECOND EDITION

K. S. Spiegler
College of Engineering
University of California, Berkeley

PLENUM PRESS • NEW YORK AND LONDON

189030

Library of Congress Cataloging in Publication Data

Spiegler, K   S
  Salt-water purification.

  Includes bibliographical references and index.
  1. Saline water conversion. I. Title.
TD479.S63 1977   628.1'67   77-21330
ISBN 0-306-31030-9

© 1962, 1977 Plenum Press, New York
A Division of Plenum Publishing Corporation
227 West 17th Street, New York, N.Y. 10011

All rights reserved

No part of this book may be reproduced, stored in a retrieval system, or transmitted, in any form or by any means, electronic, mechanical, photocopying, microfilming, recording, or otherwise, without written permission from the Publisher

Printed in the United States of America

Dedicated to the memory of my parents

PREFACE

This introductory book is for newcomers to the field of desalting. As a result of the development of arid regions, and also in the wake of intensive use of water in urban areas all over the world, fresh water is frequently not available in the quantities desired. In the last 20 years, great strides have been made in the development of the science and technology of water desalting. Statesmen, economists, scientists, and engineers must often make decisions relating to this field. Quite often such decisions involve the disposition of considerable amounts of energy and investment funds. It is hoped that this book will provide newcomers with sufficient background to appreciate the fundamentals involved, to understand the jargon of the "insiders," to guide them to the rich original multidisciplinary literature of this field, and to facilitate the first stages of further reading.

Conversion of salt water into fresh water requires useful energy, i.e., energy that could otherwise be used for mechanical or electrical work. Therefore, the problem of water desalting is intimately linked to power supply. A second, equally important factor in the economic balance is the availability and cost of equipment. The production of energy-efficient and reliable desalting plants requires considerable engineering skill and usually a variety of fairly sophisticated control devices. Because water is customarily needed and consumed in large quantities, the establishment of desalting plants involves well-developed technology and the use of large amounts of reliable and relatively expensive materials, e.g., special alloys and/or polymers. For a given desired production rate, the size of the required plant can often be decreased if the power consumption is allowed to increase and *vice versa*. Therefore, in addition to civic and political considerations, the planner must often make decisions on the relative contribution of power costs and investment. These two factors vary with local conditions and with time. To strike this delicate balance, some technical knowledge about desalting processes is essential.

Since the publication of the first edition of this book in 1962, new methods of desalting, in particular reverse osmosis (hyperfiltration), have been developed from the pilot plant to the full industrial scale, distillation technology has been refined, and many large plants have been built. This second edition was prepared with these developments in mind, introducing new material and making some shifts in emphasis, while trying to maintain the basic intent of the book: to provide a simple explanation of the technical fundamentals of desalting in a volume of small size.

CONTENTS

1 • Introduction 1
2 • The Composition of Natural Salt Waters 15
3 • Power Requirements 21
4 • Scale 29
5 • Distillation Methods 47
6 • Electrodialysis 101
7 • Freezing Processes 119
8 • Ion Exchange 131
9 • Reverse Osmosis (Hyperfiltration) 141
10 • Summary and Conclusions 155
   Appendices 163
   Index 183

1

INTRODUCTION

WHERE IS SALT-WATER PURIFICATION NECESSARY?

The oceans represent the largest water reservoir of the earth. Ocean water, however, contains on the average 3½% (by weight) of dissolved salts, a concentration that makes the water unsuitable for many uses in household and industry and for the irrigation of conventional land-grown crops. Radiation from the sun causes the evaporation of enormous quantities of water from the ocean, but the salts are not volatile and do not evaporate. When the water vapor reaches cold regions in the atmosphere, it condenses and forms clouds which eventually precipitate as rain water, containing only a minute concentration of dissolved matter. Most of the rain falls directly into the ocean and is never used by human beings, but some of the rain falling on land runs off the surface, forming rivers which eventually reach the sea. Another portion of the rain percolates into soil and rock and runs off very slowly toward the ocean through permeable underground strata. In either case, the water interacts with rock and soil and thus carries soluble minerals to the ocean. Surface waters also collect soluble and insoluble organic and inorganic materials as a result of contact with civilization, vegetation, and wildlife.

We might think that the salt concentration of the oceans is continually increasing; however, much geological evidence shows that no large changes in the average composition of sea water have occurred in millions of years. Hence it must be assumed that sedimentation on the ocean shelves and the continental floor plus spray losses roughly balances the influx of minerals into the sea. There is no evidence that much water reaches the ocean from the interior of the earth or from the cosmos. Hence all the water evaporating from the sea eventually returns to it as rain, surface water, or underground flow, thus closing the gigantic distillation cycle.

Since nature very obligingly provides this water-purification mechanism, it seems reasonable that water should be collected wherever possible before reaching the ocean, instead of fresh water being produced from sea water in man-made factories. The uneven distribution of rainfall, however, with respect to region and/or season causes a dearth of fresh-water sources in some areas and an overabundance in others.

We are now witnessing a large drive to open arid areas for large-scale settlement. This trend is the result of the increase in world population, the feasibility of control of indoor climate, and various military, economic, and political factors. Moreover, water shortages appear occasionally in highly industrialized areas in zones of moderate climate with abundant rainfall. When there is no fresh water within a reasonable distance, and ocean water or brackish (salty) ground water is available, salt removal from water becomes necessary. This process is called "salt-water purification" or "salt-water conversion." Purification plants have been in operation for years in the Virgin Islands, in the Persian Gulf area (Kuwait, Saudi Arabia), in Aruba (West Indies), in Arizona, in Eilat (southern Israel), and in many other places. Sea-water distillation aboard oceangoing vessels has been standard practice for over a century and purification plants are mushrooming in many parts of the world.

According to El-Ramly and Congdon,* 1036 land-based desalting plants, each producing 100 metric tons or more, were in operation or being constructed in the world as of January 1, 1975. These plants were capable of producing about 2 million tons (528 million U.S. gallons) per day of desalted water. While about 85% of these plants used distillation, salt separation by compression of saline water through membranes ("hyperfiltration" or "reverse osmosis") has gained increased importance in recent years. For instance, a large plant planned to desalt the saline Wellton-Mohawk drain influent into the lower Colorado River

---

*For detailed references, see the lists of selected literature at the end of each chapter.

(capacity of more than 400,000 tons per day), which prospect was not included in the pre-1975 survey, will use membrane methods exclusively, and even sea-water desalination by reverse osmosis is now feasible and possibly economical.

The process of salt-water purification—the recovery of fresh water from saline water—is more radical than the softening of water, in which merely the calcium and magnesium salts are removed or exchanged. A number of conversion methods are known which have been studied from many aspects. In fact, the large-scale purification of salt water is basically not a problem of feasibility but of economics. First, all purification processes require power. There exists a theoretical minimum of power for the production of fresh water from ocean water.

This minimum power requirement is about 1 kilowatt (kw) for each hourly production rate of 1 metric ton. In other words, the theoretical minimum power for producing 1000 tons of water in a 24-hr day is about 42 kw. In practice, much more power is used. Even if power were available free of charge, a relatively large investment would be necessary to build desalination plants. When only small amounts of water for emergency use are needed, these considerations are not of overriding importance. But the investment is very large for plants providing enough water to satisfy the needs of communities in industrially developed countries. Municipal waterworks performing rather simple manipulations such as filtration at low pressures for sanitary reasons and/or softening represent appreciable investments because the handling of large amounts of water necessitates large pumps, pipes, and other equipment. Salt-water purification processes call for even more complex equipment. It is very unlikely that salt removal from sea water will ever require less investment than the simple processes of filtration and softening of fresh water. (In fact, filtration and/or softening is sometimes required to *pretreat* the water which is to be desalted.) The main economic difficulty thus lies in the discrepancy between the customary low cost of fresh water and the need to subject salt water to a more or less intricate chemical manufacturing process to extract fresh water from it. After all, industry knows no manufactured chemical that sells for as little as 35¢ per ton. Yet many large-scale users of water in industry and agriculture would not pay this price for water under normal conditions.

The situation is entirely different where relatively small amounts of water are needed under special conditions. Sea-water distillation aboard oceangoing vessels proves economical because the combined weight of the fuel and evaporation plant is usually considerably less than the weight of the fresh water which the boat would have to carry in the absence of the purification plant. When water is needed for advance parties in isolated locations, e.g., radar towers or oilfields in desert regions, salt-water purification is often much more economical and safer than the transportation of fresh water to these sites.

To gauge the magnitude of the conversion problem for a hypothetical new community established in an area in which only salt water is available, it is necessary (1) to examine the present rates of fresh-water use, (2) to define the maximum salt concentrations which can be tolerated for each of these uses, and (3) to investigate the possibility of partial substitution of salt water for fresh water.

AMOUNT OF WATER USED

If we divide the amount of water used in a given location by the number of human beings living there, we obtain the amount of water which is directly and indirectly at the individual's disposal. This amount is called the *water index*. It includes water reused with or without purification. "Water used directly" refers to water withdrawn from rivers, lakes, springs, cisterns, or wells and can often be determined simply by reading the water meter of a person's house or apartment. The difference between this reading and the water index is more pronounced the higher the level of industrialization of a country. The desert nomad uses only a few tens of liters of water daily for himself and his cattle, and, since he buys few products and services from the outside, this small amount of water is not very different from his water index. The situation is entirely different in a highly industrialized society. The very approximate figures in Table 1.1 illustrate this point. The figures for industrial use include the large amounts of cooling water used in steam-driven power plants which can use *saline* water for this purpose. (Water that

TABLE 1.1. Amount of Water Used in the United States in 1960 (daily averages)[a]

For major industries	570 million m^3/day
For irrigation	510 million m^3/day
Other uses (household, animal watering, etc.)	100 million m^3/day
Total water used	1180 million m^3/day
Number of persons using it	180 million
Water index = $\frac{1180}{180}$	6.6 m^3/person and day
Average use of water in U.S. apartments	0.2 m^3/person and day
Average use of water in U.S. family houses	0.5 m^3/person and day
Average use of water in U.S. office buildings	0.12 m^3/person and day
Average use of water in U.S. hospitals	1.0 m^3/bed and day

[a] For conversion to other units, see Appendix 9A.

passes through the turbines of hydroelectric power plants, all of which is recovered practically without change of salinity and with only minor change of temperature, is not included in Table 1.1.)

Since 1960 the use of fresh water has been growing. The figures in Table 1.1 are presented only for illustration of the approximate consumptions for different uses and their *relative* importance. In 1970, about 1420 million cubic meters* (375 billion U.S. gallons per day) were used in the United States. This figure represents a water index of 6.9 m^3 per day. About two-thirds of this water consumption came from fresh surface water, 18% from fresh ground water, 14% from saline surface water, and less than 1% from reclaimed sewage.

We can see that the amount of water going directly to the citizen of an industrial society and of which he is aware when paying his water bill represents only a small fraction of his share in the nation's water balance. Most of the water goes to him indirectly, either from a primary water source or as reused water, for the production of power and manufactured goods and as irrigation water for food and industrial crops. The complete water index, and not merely the amount of household

*One cubic meter (m^3) of pure water weighs 1 metric ton.

water used per person, provides the basis for estimates of the water needs for a self-supporting society under normal peacetime conditions. It is obvious that the water index varies with the climate and the degree and type of industrial activity of the region.

SALINITY SPECIFICATIONS

For Human Consumption

Sea water cannot be substituted for fresh water in the diet of human beings or cattle. The maximum permissible concentration of salts in *drinking water* depends on the type of salt, the total daily water consumption, and individual factors. The United States Public Health Service recommends that the salinity of drinking water be less than 500 parts per million (ppm) and sets 1000 ppm as the highest limit (1000 ppm represents 1 g of salts per kilogram of water or 2 lb per short ton). This recommendation, however, is no more than a rule of thumb. The drinking of only a few liters per day of some brackish ground waters containing high concentrations of the sulfates of sodium and/or magnesium often causes diarrhea. On the other hand, certain North African tribes drink water containing more than 2500 ppm of dissolved salts without apparent ill effect. If the climate is hot and if the salt in this water is mainly sodium chloride, this salinity can actually be beneficial, depending on the state of health of the consumer; however, it amounts to only about one-fourteenth of the salinity of average sea water.

Since the amount of drinking water needed by humans is relatively small, varying between 2 and 8 liters per person daily, depending on climate and occupation, provision of drinking water by conversion of salt water presents no serious economic problem. For instance, a compression still, such as used by the United States Navy during World War II, provides about 7000 liters distilled water from sea water per day, using about 40 liters of gasoline. The unit with auxiliaries occupies less volume than the amount of water it produces in 1 day. The price of such units is comparable to that of an average automobile. For extreme

INTRODUCTION

emergencies, e.g., for use on liferafts, ion-exchange desalting kits producing six times their own weight of drinking water from sea water and portable solar stills have been developed.

For Industry

Purity specifications for industrial waters vary widely and depend entirely on their intended use. Much of the water used in the textile, leather, paper, chemical, and food industries has to be not only of low salinity but also of very low hardness. In general, corrosion protection in a salt-water environment is more expensive than in fresh water. This consideration rules out sea water for many applications. The large amount of salt residue left after evaporation further limits it for industrial use. For cooling water in industrial condensers, highly saline water is often acceptable. Indeed, very large amounts of sea water are being used for this purpose.

For Agriculture

Since rain, the natural irrigation water, is almost free of salts, it is perhaps not surprising that most land plants are adapted to water of low salinity. Unlike much of the vegetation (e.g., algae) living in the ocean, most land plants have not developed mechanisms to continuously maintain the relatively low salinity of their cell fluids, saps, and juices in a highly saline environment. Yields decrease with increasing salinity of the irrigation water, except at very low salinities where there may be an initial slight increase in yield with salinity. Most plants will not grow at all when irrigated with sea water. The specifications for irrigation water depend not only on the nature of the plants and dissolved salts but also on the type of soil. Plants in sandy, porous (light) soils tolerate higher water salinities than those in dense soils where drainage is poor. The method of irrigation is also important. "Drip irrigation," in which water is released directly into the soil through controlled leaks from irrigation pipes, can utilize water of higher salinity than overhead-sprinkler irrigation, which causes considerable evaporation before the water intended for irrigation reaches the plant roots. There is no doubt that the tolerance limits for some crops can be stretched by systematic agricultural research, which must also consider that—seemingly paradoxically—the addition of certain salts, e.g., potassium sulfate, to the irrigation water can raise the tolerance of the plants to common salt. For the bulk of the agricultural crops, however, irrigation with water containing more than 2000 ppm dissolved salts causes appreciable decrease of yield. The exact tolerance limits have to be determined for each combination of plant, soil, climate, and salt-water type, either by experiments or from records of agricultural research stations.

For instance, by proper matching of plant strain, soil type, and irrigation method (drip type, with water-soluble fertilizer added to the irrigation water) it proves possible to grow tomatoes on the western side of the Arava Valley (between the Dead Sea and Red Sea) in January using irrigation water which contains more than 2000 ppm of dissolved salts (before addition of fertilizer); but even this salinity is below *one-tenth* of the concentration of salts in sea water. Special strains of barley prove tolerant to very high salinity.

Enormous amounts of water are needed for irrigation of arid areas. Hence at first thought this looks like the largest potential source of application of sea-water purification. The water needs of land under intensive cultivation are about 1 m^3 (1 metric ton) per square meter per year. This amounts to 1,000,000 m^3 per square kilometer of cultivated land. Although this figure is reduced by the amount of usable rainfall, it is clear that the desalting cost would have to be very low if desalted water were to be used for irrigation. It must be borne in mind that the weight of the marketable agricultural product is very much less than the weight of the irrigation water used for its production, generally less than 1%. Instead of comparing the cost of sea-water desalting with the cost of bringing water from a region where it is plentiful, we have to weigh water costs against the cost of importing the agricultural product. Of course, the decision will depend on more than the weight ratio and freight costs, e.g., relative yields per unit area, cost of labor, and, last but not the least important, national and political factors.

INTRODUCTION

CAN SALT WATER PARTIALLY REPLACE FRESH WATER?

Since our industrial civilization is based on the premise that fresh water is a very inexpensive commodity, large industrial centers are almost always found close to abundant sources of fresh water. But as these industrial centers expand, the difference between water supply and demand narrows. Often purification plants have to be built for sewage discharged into rivers to ensure a supply of usable water for communities farther downstream. As water costs in these areas rise, we might speculate on what changes would occur in our accustomed way of life and in our industry if partial substitutes had to be found. Would many communities located on the ocean shore have dual water systems—one for fresh water, the other for sea water, using the latter for such purposes as toilet flushing? (At present, urban dual supply systems are considered uneconomical under most circumstances.) Would industry radically change its production processes and go to the same lengths to design for water economy as it already does for fuel economy? Surely a certain percentage of fresh water could be replaced and saved. Some fresh water needs are purely psychological. Consider, for instance, the domestic water consumption of the comparable cities of Haifa and Tel-Aviv, Israel, before World War II. At that time, water meters existed in the apartments in Haifa, while in Tel-Aviv a flat monthly rate was charged, depending only on the size of the apartment. The domestic per capita consumption in Haifa was less than half that in Tel-Aviv! With increasing water prices there are apt to be changes in the water regimen, but any such changes will be slow because of the tremendous investment of industrial society in a way of life based on the availability of plenty of fresh water.

But even under these circumstances there is no adequate substitute for fresh water in sight. Since the plans for development of many arid countries as well as the mild, but increasing, water shortages in industrial regions necessitate additional fresh-water supplies, there has been a steep increase of activity in salt-water purification research, development, and plant construction in the last two decades.

SOURCES OF DETAILED INFORMATION

Since salt-water purification affects a large number of countries, the United Nations stimulated an international exchange of information on the subject at the start of this two-decade period which was critical for the development of desalination science and technology.

While excellence knows no national boundaries and contributions to the science and technology of desalting have been made in many countries, by far the largest research and development effort has been made under the sponsorship of the United States Department of the Interior through its Office of Saline Water (OSW), now part of the Office of Water Research and Technology. Hundreds of detailed reports, listed in the Annual Reports of OSW, which existed as an independent unit until 1974, have been published and are available at nominal cost from the Superintendent of Documents, U.S. Government Printing Office, Washington, D.C. 20402. Abstracts of articles dealing with desalination are commercially available (see the list of selected literature at the end of this chapter).

Original articles on desalting can be found in various journals, particularly in the journal *Desalination* (Elsevier, Amsterdam), which specializes in this subject.

Before delving into specifics, readers interested in a particular aspect of desalting might first wish to consult one of the existing *summary articles* which distill the essence of many cited references. An extensive summary by A. A. and E. A. Delyannis was published in 1974. *Principles of Desalination* (Academic Press, New York) is a collection of summary articles with extensive lists of references covering many aspects of desalting science and technology.

I have added lists of recommended additional reading at the end of each chapter of this book. While these lists are not meant to be exhaustive, it is hoped that the references cited, which can be found in most libraries of major universities, will guide the reader on the next step of his study.

SELECTED LITERATURE

General

Deming, H. G., *Water—The Fountain of Opportunity,* Oxford University Press, New York, 1975. Original manuscript revised and updated by W. Sherman Gillam and W. H. McCoy.

Reviews

Delyannis, A. A., and Delyannis, E. A., *Water Desalting: Gmelin's Handbook of Inorganic Chemistry,* supplementary volume (Anhangband) to *Oxygen,* Springer-Verlag, New York, 1974. A very comprehensive compilation of literature data.
Spiegler, K. S., and Laird, A. D. K., eds., *Principles of Desalination,* 2nd ed., Academic Press, New York, 1977. This monograph contains comprehensive chapters, most with substantial literature reviews (to 1975), on the major desalting methods, on desalting energetics, and on the cost of conventional water supply.
Apelcin, I., and Klyachko, V., *Opresnenie Vodi [Water Desalination],* Moscow, 1968. In Russian, 200 pp., paperback.

Abstracts

Desalination Abstracts, National Center for Scientific and Technological Information, P.O. Box 20125, Tel Aviv, Israel. An extensive international quarterly compilation listing books, journal articles, and patents by subject; started in 1965.
Desalination Abstracts, P.O. Box 1199, Omonia, Athens, Greece. Started in 1969 and appears every second month.
Selected Water Resources Abstracts, Office of Water Research and Technology, U.S. Department of the Interior, for sale by the National Technical Information Service, U.S. Department of Commerce, 5285 Fort Royal Road, Springfield, Va. 22161. Appears semimonthly, covering many aspects of water resources. Index lists articles on "Desalination."
Bulletin de l'association française pour l'étude des eaux, Association française pour l'étude des eaux, 21 rue de Madrid, 75008 Paris, France. In French; appears monthly and contains selected abstracts on many water-resource topics, including desalination, arranged in order of degree of specialization.

Water Resources Research Catalog, Office of Water Research and Technology, U.S. Department of the Interior, Washington, D.C., for sale by Superintendent of Documents, Washington, D.C. 20402. Lists ongoing research projects in water technology in general; desalting projects are found in the index under *Chemical Engineering:* "Desalination" and *Waste Control and Water Quality:* "Industrial Wastes."

Symposia Proceedings

National Academy of Sciences–National Research Council, *Proceedings of the Symposium on Saline Water Conversion 1957,* Publ. 568, Washington, D.C., 1958.
American Chemical Society, *Saline Water Conversion* (Proc. 1960 Symp.), *Advances in Chemistry Series,* No. 27, Washington, D.C., 1960.
American Chemical Society, *Saline Water Conversion II* (Proc. 1962 Symp.), *Advances in Chemistry Series,* No. 38, Washington, D.C., 1963.
Proceedings of the First International Symposium on Water Desalination, 1965, 3 vols., for sale by Superintendent of Documents, Washington, D.C. 20402.
Proceedings of the Third [Fourth, Fifth] International Symposium on Fresh Water from the Sea (1970, Dubrovnik, 1973, Heidelberg, and 1976, Alghero, respectively), 4 vols. each, for sale by A. A. Delyannis and A. E. Delyannis, Tsaldari St. 34, Athens-Amaroussion, Greece.

Bibliography

Ten volumes covering period to 1968. For sale by A. A. Delyannis and A. E. Delyannis (address given above).

Newsletters

Water Desalination Report, for sale by Richard Arlen Smith, Publisher-Editor, P.O. Box 35-K, Tracey's Landing, Md. 20869. Appears weekly.
Water Newsletter, for sale by Water Information Center, 14 Vanderventer Avenue, Port Washington, N.Y. 11050. Covers many fields, including selected news about desalination.

Current Journals Featuring Original Articles

Desalination, Elsevier Scientific Publishing Co., P.O. Box 330, Amsterdam, The Netherlands. Features primarily articles on the science and technology of salt-water purification.

Journal of the National Water-Supply Improvement Association, Patricia Burke, Managing Editor, c/o AVCO Systems Division, 201 Lowell St., Wilmington, Mass. 01887. Features primarily articles on the management and economics of desalting piants, edited by N. A. El-Ramly.

Desalting Plant Inventories

El-Ramly, N. A., and Congdon, C. F., *Desalting Plants Inventory Report No. 5,* 1975, Office of Water Research and Technology, U.S. Department of the Interior, Washington, D.C. 20240.

Second United Nations Desalination Plant Operation Survey, 1973, Resources and Transport Division, Centre for Economic and Social Information, New York or Geneva, Sales No. E. 73.II.A.10.

Water Use

Picton, Walter L., *Water Use in the United States 1900–1980,* U.S. Department of Commerce Report, 1960, for sale by Superintendent of Documents, Washington, D.C. 20402.

Murray, C. R., and Reeves, E. B., *Estimated Use of Water in the United States in 1970,* U.S. Geological Survey Circular 676, Washington, D.C. 20242.

Use of Saline Water in Agriculture

Proceedings of the International Symposium on Brackish Water as a Factor in Development, A. S. Issar, ed., Ben-Gurion University of the Negev, Beer Sheva, Israel, 1975.

Boyko, H., and Boyko, E., *Proceedings of the International Congress of Bioclimatology and Biometeorology,* Vol. 3, Part 2, Section B1, Leiden, The Netherlands, March 1959.

Scientific American, Vol. 235 (2), p. 44D, August 1976. Brief description of work by E. Epstein and J. Norlyn on irrigation of a special strain of barley with sea water.

Drip/Trickle Irrigation, a journal published every second month by the International Drip-Irrigation Association, 17068 Glentana St., Covina, Calif. 91722.

2

THE COMPOSITION OF NATURAL SALT WATERS

Before dealing with the removal of salts, it is necessary to know exactly their nature and concentration. Sea water contains all known elements, but most of them are present in only minor concentrations.

The exact concentrations of these minor constituents are not accurately known. Many older determinations of their concentrations are apparently erroneous. The exact magnitudes of the concentrations of some of the minor constituents (e.g., phosphates and nitrates) are often very important for the flora and fauna of the oceans, and for special problems, such as the recovery of gold from sea water, which has been attempted and given very poor yields, possibly because of overoptimistic analytical results. For the purpose of sea-water desalting, however, it is well to focus on the major constituents listed in Table 2.1. Minor constituents of salt water cannot always be disregarded, however. Thus a certain brackish water containing a minor concentration of barium ions was found to deposit insoluble barium sulfate deposits in the electrodialysis units in which it was treated.

In Table 2.1 the concentrations of the *ions* rather than of the salts are given, since the salts are almost completely dissociated in sea water. We can see that sodium and chloride ions are present in the largest

TABLE 2.1. Major Constituents of Sea Water (in parts per million)[a]

Sodium (Na^+)	10,561
Magnesium (Mg^{2+})	1,272
Calcium (Ca^{2+})	400
Potassium (K^+)	380
Chloride (Cl^-)	18,980
Sulfate (SO_4^{2-})	2,649
Bicarbonate (HCO_3^-)	142
Bromide (Br^-)	65
Other solids	34
Total dissolved solids	34,483
Specific gravity (20°C)	1.0243
Water (balance)	965,517

[a] Chlorinity 19.000, salinity 34.325.

concentration, but sea water is by no means a solution of pure sodium chloride (common salt). Calcium and magnesium bicarbonate and sulfate are the troublemakers in most desalting processes. They yield various insoluble deposits (e.g., calcium carbonate, magnesium oxide, and calcium sulfate) which, when unchecked, impair the proper operation of many types of desalting installations.

The total salt concentration of sea water is expressed in terms of *salinity,* which is approximately equal to the total amount of dry solids (in grams) per kilogram of sea water, i.e., parts per thousand, or the *chlorinity,* which is very nearly the concentration of chloride ions (also in parts per thousand).* Salinity and chlorinity vary with depth and geographical position. In the open ocean the salinity lies between 33.6% and 36.8%, but in more isolated basin areas extreme deviations can occur. Thus in large portions of the Baltic Sea, where precipitation and inflow of fresh water outweigh evaporation, the surface salinity drops to less than 7 parts per thousand. On the other hand, in the Red Sea, salinities up to 41 parts per thousand have been measured. Thus the highest salt concentrations are found just where the need for desalting is greatest.

In spite of these considerable differences of salinity, the *relative* abundance of the major constituents of sea water is about the same

---

*Exact definitions are listed in Appendix 7A.

everywhere. When high rainfall dilutes the surface layers so that the sodium concentration drops one-half, all the other major constituent concentrations are halved also.

Ocean water is not the only highly saline surface water; some large lakes in arid regions, without drainage, contain saline water because the water fed into them by the rivers evaporates, whereas the salts remain. A few examples are listed in Table 2.2.

Finally, there exist many *saline ground waters* in arid regions. In general, the salinity of ground waters increases with depth, although there are exceptions. In drilling for petroleum we usually penetrate porous rocks containing salt water much more highly concentrated than sea water. These waters are generally too deep beneath the surface to warrant utilization. The reasons for the high salinity of the brackish ground waters in arid regions are rarely accurately known. Collection of airborne salts and high evaporation rates coupled with poor drainage certainly cause higher salt concentrations than in nonarid regions. Migration of some of the deeper and still more saline ground waters through cracks may sometimes be a major factor in promoting high salinity. When planning a purification plant for brackish ground water, it is very important to make a thorough hydrological study of the particular underground strata to make sure that a sufficient supply of water of a given quality can be maintained. Frequently no past experience on this

TABLE 2.2. Concentration of Salts in Different Salt Waters (parts per million)

	Dead Sea (northern part, 100 meters below surface)	Great Salt Lake, Utah	Caspian Sea (V. B. Chernozubov *et al.*)
Sodium	32,000	67,300	3,350
Magnesium	35,700	5,600	800
Calcium	12,700	300	360
Potassium	6,400	3,400	70
Chloride	178,600	112,900	5,540
Sulfate	400	13,600	3,170
Bicarbonate and carbonate	Trace	200	2,385
Bromide	5,200	Trace	

point is available. Since the well yields only salt water, it has often not been used extensively and no one knows how the salinity of the water is going to change when major amounts of brackish water are withdrawn. In this respect, sea water often offers the advantage of a continuous supply of almost unchanging quality.

The composition and total concentration of salts in brackish ground waters vary greatly with the location. The major components are the same as in sea water but their relative abundances are often quite different. Table 2.3 lists the composition of some selected brackish waters and also of some fresh-water supplies.

The table presents also the composition of a brackish irrigation drain water (Wellton-Mohawk Drain) which flows into the lower Colorado River at an average rate of about 0.5 million tons per day. Such brackish drain waters are quite common in intensely irrigated areas, and they cannot be avoided since some drainage must always be provided so as to remove excessive salinity from the root zone of the irrigated crops. To reduce the salinity of lower Colorado River water flowing into Mexico, the U.S. Department of the Interior is in the process of supervising the building of a desalting plant (based on membrane methods) to reduce the Wellton-Mohawk inflow into the Colorado River to tolerable limits. To dispose of the brine (saline concentrate) produced, a special canal to the Gulf of California (in Mexico) is to be built. Thus a large portion of the salts is to be eventually sent to the ocean.

TABLE 2.3. Composition of Some Water Supplies[a]

	Na^+	Mg^{2+}	Ca^{2+}	Cl^-	SO_4^{2-}	HCO_3^-
Lake Erie at Buffalo, N.Y.	7	7	31	9	13	114
Shenandoah River at Mileville, W. Va.	7	8	32	3	6	132
Brackish Well, Salt Lake City, Utah	685	139	266	1940	33	67
Brackish Asi Spring, Beth Shean, Israel	434	71	186	916	113	334
Wellton-Mohawk Drain (1972)	790	80	330	1360	760	360

[a] All figures are in parts per million. Minor constituents, e.g., nitrate, iron, silica are not listed.

Surface waters have generally relatively low salinity, composed mostly of salts of calcium and magnesium. These salts cause the *hardness* of the surface waters. Sea water contains an even higher concentration of these salts and is therefore very hard. In addition, however, it contains such large amounts of sodium chloride that the weight of sodium exceeds that of calcium and magnesium sixfold. Hence softening alone is used to upgrade the quality of many surface waters, while seawater purification calls for the more radical treatment of almost complete salt removal.

SELECTED LITERATURE

Ocean Salinity

Sverdrup, H. U., *Oceanography for Meteorologists*, George Allen and Unwin, London, 1945.

Sverdrup, H. U., Johnson, M. W., and Fleming, R. H., *The Oceans*, Prentice-Hall, New York, 1942. Baltic Sea, see p. 657.

Riley, J. P., and Skirrow, G., *Chemical Oceanography*, 6 vols., Academic Press, New York (vols. 1 and 2 of 2nd ed.), 1975.

Caspian Sea

Chernozubov, V. B., Zaostrovskii, F. P., Shatsillo, V. G., Golub, S. I., Novikov, E. P., and Tkach, V. I., "Prevention of Scale Formation in Distillation Desalination Plants by Means of Seeding," *Proceedings of the First International Symposium on Water Desalination*, Vol. 2, Washington, D.C., 1965, p. 539, for sale by Superintendent of Documents, Washington, D.C. 20402.

Baltic Sea

Ahlnäs, K., "Variations in Salinity at Utö 1911–1961," *Geophysica (Helsinki)* **8(2)**:135 (1962).

Dead Sea

Bentor, Y. K., "Some Geochemical Aspects of the Dead Sea and the Question of its Age," *Geochim. Cosmochim. Acta* **25**:239 (1961).

Wellton-Mohawk Drain

Preliminary Sizing Study, Yuma Desalting Plant, Arizona, Table 13, U.S. Department of the Interior, Bureau of Reclamation, Denver, Colo., p. 15a.

Contribution of Ocean Spray to the Salinity of Inland Waters

Löwengart, S., "Airborne Salts—The Major Source of the Salinity of Water in Israel," *Bull. Res. Council Israel* **10G:**183 (1961).

sea water and the latter is subsequently rejected. In practice, we carry the process further in order to avoid pumping too large amounts of sea water. For recovering 30, 50, 67, and 75% of the water, the minimum work requirements are 0.84, 0.97, 1.16, and 1.30 kwhr/m^3, respectively, as discussed and listed in Appendix 1A. The work requirements for water of lower salinity than sea water are less and can be calculated from the formulae in the same appendix.

All salt-water purification processes split the raw water into (1) a more concentrated portion and (2) a more dilute portion or completely pure water. Therefore, it is not surprising that work must be invested in the process. Nature tends to equalize differences. When a salt solution and fresh water are placed into a container, they will invariably mix until the concentration in the resulting solution is uniform. Nobody has ever observed a solution separating spontaneously into a more concentrated and a more dilute portion. In this respect, the problem is similar to that in refrigeration. Any household refrigerator cools the inner space and simultaneously heats the outer space. This process, too, requires the expenditure of work (by compressor) or (as in some compressorless refrigerators) the flow of heat from high to low temperature—a process which could, in principle, be used to produce work.

NONIDEAL PROCESSES

The *minimum power requirement* refers to ideal processes with absolutely no power losses. In practice, it is impossible to reach this goal. All practical processes involve friction, imperfection of mechanical and/ or electrical equipment, pumping of product and brine concentrate, and other expenditures of power which have not been taken into account in the calculation presented above.

In general, a process approaches the theoretical minimum power requirement only when it is carried out extremely slowly. This would require huge plants and completely new design concepts for equipment to work efficiently at such low rates. It would be quite impractical to plan for power economies close to the theoretical minimum. When the energy is supplied as heat, we must consider that only a portion of the heat can be converted into work. This portion depends on the process. Thus the conversion of heat into electrical power in a modern plant is carried out at an efficiency of 30–40%.

In spite of these qualifications, the minimum work consumption figures are of value to students of the salt-water conversion question. They indicate that limit which nature itself sets to the solution of the problem. They are independent of the degree of perfection of the available machinery, which is likely to vary with the ages.

To illustrate the significance of nonideal conditions, let us consider a process in which energy is wasted by discharging brine and distillate at a higher temperature than the incoming sea water. If this temperature rise is as little as 3°C, the waste of heat is 3000 kcal/m^3 of brine and distilled water combined. If half of the incoming sea water is converted to fresh water, the heat wasted is 6000 kcal/m^3 of product. Compare this to the theoretical minimum heat requirement for the process. At 25°C the work requirement is 0.97 kwhr/m^3 (Appendix 1A), thermally equivalent to 835 kcal/m^3. Hence the heat loss through this small temperature increase amounts to over seven times the available energy necessary for an ideally reversible desalination plant. It follows that processes using several times the minimum available energy are the best that we can expect. Of course, it is not fair to compare work and heat directly because the latter cannot be entirely converted into the first (Appendix 2A). Even at a conversion efficiency of heat to work of 40%, however, the power loss is still about three times higher than the theoretical work requirement.

Pumping requirements are likely to add less to the power bill, but they are not insignificant either. Compare the theoretical minimum work of 0.97 kwhr/m^3 product to the pumping costs. Since 0.97 kwhr equals 356 ton-m, the minimum work is equal to the work necessary to lift 1 ton of sea water 356 m.* If the pumps in the plant have to work against a head of 40 m and if the pump efficiency is 67%, the pumping work adds another 60 ton-m per ton of sea water treated. If 1 ton of fresh water is produced from each 2 tons of sea water, this amounts to 120 m per ton of product or about an additional one-third of the theoretical minimum work for desalting at 50% recovery.

*All references are to *metric* tons unless stated otherwise.

SELECTED LITERATURE

General Energetics, "Useful" Energy (Availability, Exergy)

Hatsopoulos, G. N., and Keenan, J. H., *Principles of General Thermodynamics,* Wiley, New York, 1965. In English.

Chartier, P., Gross, M., and Spiegler, K. S., *Applications de la thermodynamique du non-équilibre: Bases d'énergétique pratique,* Hermann, Paris, 1975. In French.

Baehr, H. D., *Thermodynamik,* 2nd ed., Springer-Verlag, Berlin, 1966. In German.

Energetics of Water Desalting

Tribus, M., "Thermo-economic Considerations in the Preparation of Fresh Water from Sea Water," *Dechema Monogr.* **47**:43 (1962).

Bromley, L. A., Singh D., Ray, P., Sridhar, S., and Read, S. M., "Thermodynamic Properties of Sea-Salt Solutions," *Am. Inst. Chem. Eng. J.* **20**:326 (1974).

Gilliland, E. R., "Fresh Water for the Future," *Ind. Eng. Chem.* **47**:2410 (1955).

4

SCALE

Scale is a deposit of minerals which forms on solid surfaces of industrial equipment. It is of particular importance in salt-water purification, since it appears in almost all the processes now in industrial use. In distillation processes it can form on the evaporator tube surfaces and in membrane methods on the surfaces of the membranes which act as ionic sieves. The presence of scale leads invariably to operating difficulties and/or loss of efficiency. In distillation, scale reduces the rate of heat transfer through the affected surfaces. In electrodialysis and reverse osmosis, it plugs the membranes; it increases the electrical or the hydraulic resistance and the power consumption. Therefore, scale prevention is an integral part of the design, operation, and cost of these processes. Pretreatment of saline water prior to desalting proper often proves necessary, and frequently one of the purposes of the pretreatment is the removal or at least the reduction of the concentration of ions which can be components of scale.

The components of scale derive from the raw water, except when *corrosion* takes place simultaneously with the deposition of scale. Then corrosion products (e.g., iron oxides) may also become part of the scale. The composition of scale depends on the raw-water composition and the operating conditions. Chemical and mineralogical examinations of scale in boilers and water treatment plants have revealed a considerable number of compounds in different scales. In salt-water purification plants, scale usually contains the following chemical compounds: magnesium oxide, calcium carbonate, and calcium sulfate. These compounds are found in different solid forms, either pure or in mixtures. The formation of scale can occur where the solubility limits of these compounds are exceeded. Calcium and magnesium salts represent the *hardness* of water.

SOLUBILITY LIMITS

The solubility of the salts in sea water and brackish waters differs between wide limits. Figure 4.1 shows the solubility of three salts commonly found in these waters as a function of temperature. These data were obtained by stirring 1000 g of pure water with a large excess of the respective salts and determining the maximum amounts of salt that can be dissolved at definite temperatures.

It is seen that magnesium chloride and sodium chloride (common salt) are highly soluble. Normal sea water can be concentrated to about one-tenth of its volume before the concentration of sodium chloride reaches saturation. Unless recovery of salt or magnesium chloride is specifically desired, the raw water is usually not concentrated to such a large extent. Therefore, highly soluble salts are generally not components of the scale which forms in water-purification plants.

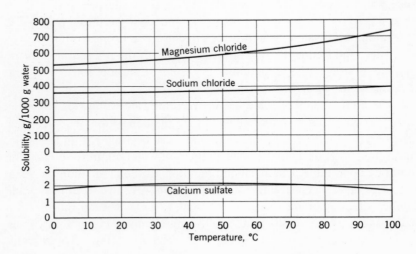

Fig. 4.1. Solubility of $MgCl_2$, NaCl, and $CaSO_4$ in water.

The solubility of both magnesium chloride and common salt *increases* with increasing temperature. Such behavior is called "normal" behavior, although the opposite is quite common. Thus the solubility of calcium sulfate, also shown in Fig. 4.1, decreases with increasing temperature above 38°C ("inverted" solubility in this temperature range). The solubility of sodium chloride increases only very little (in terms of percentage) with increasing temperature.

Each of the solubility curves of Fig. 4.1 refers to a solution of a single salt in water, but most natural salt waters contain a variety of salts. The presence of other salts often exerts a strong influence on the solubility. Thus calcium sulfate and calcium carbonate are more soluble but magnesium chloride is less soluble in concentrated sodium chloride solutions than in pure water. The presence of magnesium chloride increases the solubility of calcium sulfate appreciably. When determining solubility limits, it is necessary to make allowance for the presence of *all* dissolved salts.

When pure water is separated from salt water, the concentration of salts builds up and eventually reaches saturation with respect to one or more salts. In addition, when a raw water saturated with a salt of inverted solubility curve, e.g., calcium sulfate, is heated, the solubility limit is eventually exceeded, even without boiling off water, because the solubility of such salts at high temperatures is less than at room temperature. In both cases, supersaturated solutions are obtained, i.e., solutions containing more salt than the solubility limit. It is important that, although supersaturated solutions are unstable, solid salts do not necessarily deposit immediately from these solutions. As soon as these solutions are brought into contact with a crystal of the respective solid material, however, all the dissolved solid they contain, over and above saturation, precipitates. It has been found that not only crystals of the salt in supersaturated solution but also many other solids can act as centers of crystallization.

For many years the formation of scale in distillation processes has been explained merely by the inverted solubility curve of the common scale-forming minerals. It was reasoned that on the surface of the heating tubes the solubility of these materials is lower than in the bulk of the solution. Hence supersaturation is reached there first and is immediately followed by deposition of solid. It is now recognized that supersaturation is only one of the prerequisites of scale deposition. Unless the solution is highly supersaturated, centers (nuclei) of crystallization must also be present. If such centers exist on the heating surface, scale will deposit on it, but, if they are provided in another region of the equipment, precipitation of solid from supersaturated solution will occur in that region.

To attack the scale problem for a given raw water it is pertinent to ask first under what conditions of temperature and concentration the raw water becomes unstable, i.e., potentially scale forming. It is then often possible to combat scale by (1) designing the process such that these conditions never prevail anywhere in the desalting unit or (2) by controlling the degree of supersaturation so that deposition of solids occurs only after the water has left the unit.

Scale problems must be considered individually, i.e., in relation (1) to the raw-water composition and (2) to the desalting process proposed. Basic data on the stability limits of solutions of scale-forming salts have been determined in the classical work of W. F. Langelier and his collaborators at the University of California, Berkeley, and are discussed briefly in the following paragraphs. Although this work was primarily concerned with scale formation in distillation plants, many of the findings can also be applied to other treatment methods.

A schematic representation of scale formation in a distillation unit is shown in Fig. 4.2. The raw water flows through tubes heated by steam. Scale may then form within the tubes. If the tubes were heated by direct flame, the danger of scale formation would be much larger because the temperature of the tubes and the degree of supersaturation of materials with inverted solubility curves would be very high, and hence these materials would deposit rapidly on the hot tubes. Moreover, the high temperatures would bake the scale into a hard cake which is very difficult to remove. This does not occur in the steam-producing boiler because in it steam is generated from soft water which contains no scale-forming salts. The condensate returns to the boiler and is used again for steam generation so that no potential scale-forming materials ever enter the boiler. The small amounts of make-up water which replace inevitable losses from the closed steam-condensate loop of the boiler are either softened or completely desalted.

SCALE

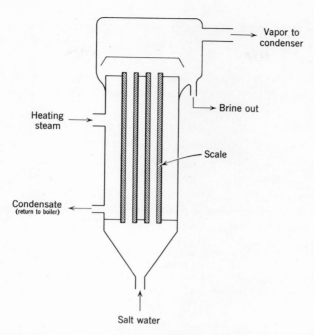

Fig. 4.2. Scale formation in a still. Steam heats the salt water flowing upward through the tubes; water boils and vapor passes on to condenser. Scale forms on inner surfaces of tubes, reduces their diameter, and slows down heat flow through tube walls. Flow of salt water can also be reversed.

CALCIUM SULFATE

The solubility of calcium sulfate is determined by the product of the molar concentrations C_{Ca} and C_{SO_4} of calcium and sulfate ions, respectively. These two concentrations are not necessarily equal. For instance, sea water contains magnesium sulfate in addition to calcium sulfate and the molar concentration of sulfate is almost three times higher than that of calcium. The product $C_{Ca} \times C_{SO_4}$ is called the *ionic product* of calcium sulfate in the respective solutions. As pure water is removed from a solution, the concentrations of the ions and the ionic product increase. Figure 4.3 shows this increase as a function of the degree of

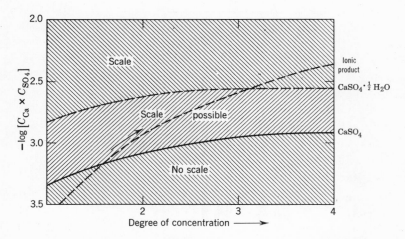

Fig. 4.3. Solubility products of two forms of calcium sulfate in concentrated seawater brines at 100°C. The solubility of the hemihydrate, $CaSO_4 \cdot \frac{1}{2} H_2O$ is higher than that of anhydrite, $CaSO_4$. Broken line shows increase of the ionic product $C_{Ca} \times C_{SO_4}$, when sea water is being concentrated. $-\log[C_{Ca} \times C_{SO_4}] = 3$ corresponds to solubility product of 10^{-3} or 4.4 g $CaSO_4$ per liter sea water.

concentration, starting with average sea water (broken line). Degrees of concentration 2, 3, and 4 refer to evaporation of one-half, two-thirds, and three-fourths, respectively, of the water from average sea water. This broken line was calculated from the original concentrations of calcium and sulfate in average sea water, assuming no precipitation takes place.

By definition, a *saturated* solution of calcium sulfate has the highest value of the ionic product possible in a stable solution at the given temperature. This *maximum* value is the *solubility product*. In Fig. 4.3 the solubility product of calcium sulfate, $CaSO_4$ (anhydrite), is also plotted. It is not completely independent of the concentrations of the other ions present in sea water, e.g., sodium, magnesium, and chloride. Therefore, it changes somewhat with the degree of concentration. All points underneath this line correspond to nonsaturated solutions. These are stable and no scale deposits from them. All points above the line correspond to concentrations higher than saturation. Although these solutions are unstable and potential scale formers, they can exist in the supersaturated state for many hours.

Calcium sulfate can also appear as the solid $CaSO_4 \cdot \frac{1}{2} H_2O$ (hemihydrate or plaster of Paris) in which two molecules of calcium sulfate share a water molecule. The hemihydrate is more soluble than anhydrite, as seen from the plot of its solubility product in Fig. 4.3. Supersaturated solutions of the hemihydrate have a relatively short life before solid hemihydrate starts to deposit from them. Hence the region above the solubility product of the hemihydrate, which corresponds to these solutions, is marked as a scale region.

These findings are of decisive importance if we wish to distill sea water without deposition of calcium sulfate scale. In this process, the ionic product follows the broken line in Fig. 4.3 in the direction of the arrow. After the sea water is concentrated to about two-thirds of its original volume (degree of concentration 1.5), the ionic product equals the solubility product of *anhydrite*. Upon further evaporation, precipitation of the latter may occur, but it is safe to continue the distillation process, provided there are no seeds present to induce crystallization. Only when about two-thirds of the water has been evaporated (degree of concentration 3.0) and the ionic product begins to exceed the solubility product of the *hemihydrate* does the danger of scale deposition become serious. It is advisable to discard the brine when it reaches this concentration.

At other temperatures, stability is described by similar diagrams with somewhat different numerical values of the solubility products.

MAGNESIUM HYDROXIDE

The stability of salt waters with respect to calcium sulfate scale is almost independent of the acidity of the water, but the reverse is true for magnesium hydroxide and calcium carbonate scale. The latter two types of scale are deposited from neutral or alkaline waters. They can be readily dissolved by acids, whereas calcium sulfate cannot.

Knowledge about the solubility of magnesium hydroxide is much less accurate than about that of calcium sulfate. The solubility product of magnesium hydroxide, $C_{Mg} \times (C_{OH})^2$, at 25°C, is of the order of 10^{-11} mole^3 liter^{-3}. In other words, when the product of the square of the hydroxide ion concentration and the magnesium ion concentration exceeds this value, precipitation of magnesium hydroxide occurs.

Consider, for example, average sea water: $C_{Mg} = 0.054$ and C_{OH} about 10^{-6} moles liter^{-1}, since the pH* is about 8. Hence the ionic product, $C_{Mg} \times (C_{OH})^2$ is 5.4×10^{-14}, i.e., less than the solubility product. It follows that average sea water is not saturated with magnesium hydroxide at 25°C; however, the dissociation of water into hydrogen and hydroxyl ions increases very rapidly with increasing temperature. When sea water is evaporated at temperatures above 70°C, a point is soon reached where the ionic product exceeds the solubility product of magnesium hydroxide which then deposits. The position of the solubility limit depends on the pH of the sea water as shown in Fig. 4.4, in which saturation curves for 60°C and 100°C are plotted. Any point to the *left* of the respective line represents solutions stable with respect to magnesium hydroxide deposition, to the *right* supersaturated solutions. For instance, point P_1 represents sea water of pH 8.1 and magnesium concentration of 1300 ppm. It is seen that water of this composition is unstable at 100°C. Hence magnesium hydroxide scale is liable to deposit if the water is just heated to 100°C even without evaporation taking place.

The water described by point P_1 is stable at 60°C, but, when concentrated to 3600 ppm (point P_2) without a change in the pH, as shown by the dashed line, it reaches saturation and further concentration produces an unstable solution. In distillation processes, the concentration does not take place at constant pH because the solution loses carbon dioxide as a result of the decomposition of bicarbonate ions:

$$HCO_3^- \rightarrow CO_2 + OH^-$$

The hydroxyl concentration increases and so does the pH. Hence the composition changes along a line sloping to the right, e.g., $\overline{P_1 P_3}$, and saturation is reached sooner than if the pH were remaining constant.

The loss of carbon dioxide during evaporation is also of crucial importance for the deposition of calcium carbonate scale.

---

*pH is defined as the negative decimal logarithm of the hydrogen ion concentration, pH = $-\log C_H$. At 25°C, the ionic product of water is $C_H \times C_{OH} = 10^{-14}$.

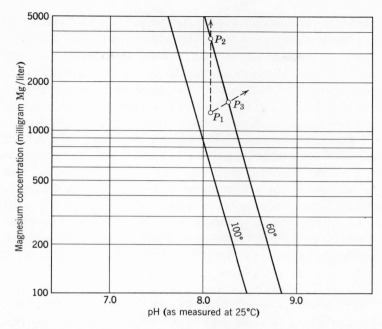

Fig. 4.4. Stability diagram for magnesium hydroxide in sea water at 60 and 100°C. Region on the right of respective line represents supersaturated, unstable solution likely to deposit magnesium hydroxide scale. Lines P_1P_2 and P_1P_3 represent evaporation of average sea water at constant pH and increasing pH, respectively. (Note that the abscissa refers to the pH measured after the solution is allowed to cool to 25°C, not to the pH prevailing at 60 or 100°C.) From Langelier *et al.*

CALCIUM CARBONATE

Calcium carbonate is very sparingly soluble in pure water. The solubility of less than 0.01 g/liter at 25°C can be greatly increased by adding acid. Even weak acids such as carbonic are effective. The latter acid is formed when carbon dioxide dissolves in water. In water subjected to carbon dioxide pressures of 1, 10, and 50 atm, the solubility rises to 0.84, 1.73, and 2.88 g/liter, respectively. The solubility decreases with increasing temperature.

Many natural waters, including sea water in many locations, are saturated with calcium carbonate. When these waters are heated, some calcium carbonate scale would deposit, even if no carbon dioxide were lost from the solution and no evaporation took place. In practice, however, the heating of salt water usually leads to loss of carbon dioxide from the water. This factor and the progressive concentration of the solution through evaporation of water are the main causes for the formation of calcium carbonate scale.

This scale, often called *soft scale*, has been known and fought since the early days of steam boiler practice. It is often the first scale to deposit from fresh water. The stability of any water with respect to this scale is determined by the solubility product, $C_{Ca} \times C_{CO_3}$, of the calcium and carbonate ion concentrations at saturation. It is relatively easy to determine the concentration of calcium ions by chemical analysis but not the concentration of carbonate ions, for the latter react partially with hydrogen ions in the water, yielding bicarbonate ions, HCO_3^-:

$$CO_3^{2-} + H^+ \rightleftharpoons HCO_3^-$$

The extent of this reaction depends on the concentration of hydrogen ions, as we see from the equilibrium equation. Most saline waters have a pH lower than 9, and in these waters the reaction goes almost to completion. In other words, the concentration of carbonate ions is very small. Instead of trying to measure it, it is convenient to determine the concentrations of bicarbonate and hydrogen ions which, at any temperature, determine unequivocally the concentration of carbonate ions.* The first can be done by a simple titration with acid and the second by colorimetric or electrometric measurement. Therefore, stability diagrams for calcium carbonate measurements are plotted in terms of these two variables, in addition to the calcium ion concentration, as shown in Fig. 4.5.

Each line divides between the stable (left) and unstable (right) region of water of a given bicarbonate ion concentration, designated here as "alkalinity" and expressed in units of "milligrams calcium carbonate per liter." In effect, the alkalinity is simply the concentration of bicar-

---

*Because the expression $C_{HCO_3}/(C_H \times C_{CO_3})$ is constant for a given temperature. For instance, at 25°C its value is 2.3×10^{10} in pure water. It is somewhat affected by the presence of other salts.

bonate ions (in milligrams per liter) multiplied by 0.82.* From these diagrams, it is possible to see if a given water is stable with respect to calcium carbonate scale at 60°C and 100°C. Consider, for instance, point P_1, Fig. 4.5a, which represents sea water of pH 8.1 and calcium concentration of 400 mg/liter. Chemical analysis has shown that the total alkalinity of this sea water is 117 mg/liter expressed as calcium carbonate. It is seen that the point lies to the right of the 117 mg/liter alkalinity line. Hence average sea water is not stable at this temperature and calcium carbonate scale is likely to deposit from it. The difference between the pH of the sample and the pH corresponding to saturation (for the same calcium concentration and total alkalinity) is termed the "saturation index" (Langelier).

Comparison of the calcium carbonate stability diagrams at 60°C and 100°C demonstrates the decrease of stability with increasing temperature. Both figures refer to sea water. For waters of different composition the stability relationships are only slightly different because the effect of the nonscaling components, e.g., sodium and potassium, is rather small, except at very high concentrations.

Applying the stability data for calcium sulfate, magnesium hydroxide, and calcium carbonate to sea water, the following conclusions may be drawn. Since sea water is practically saturated with calcium carbonate, deposition of this scale occurs already at relatively low temperatures, say 60°C. Magnesium hydroxide scale forms at higher temperatures and/or when the sea water has been concentrated to a considerable extent. The precipitation of magnesium hydroxide liberates acid and thus inhibits the precipitation of calcium carbonate:

$$Mg^{2+} + 2H_2O \rightarrow Mg(OH)_2 + 2H^+$$

The danger of calcium sulfate scale at 100°C arises only when the sea water has been concentrated to two-thirds of its volume and becomes acute only after evaporation of another third. It is therefore not surprising that in sea-water stills working at low temperature the scale (up to 60°C) consists mainly of calcium carbonate. In units operating between

---

*The exact definition of alkalinity is $[C_{HCO_3} + 2C_{CO_3}] \times 50$, but below pH 9 the carbonate ion concentration C_{CO_3} is negligible. The factor 0.82 is the ratio, 50/61, of the equivalent weights of $CaCO_3$ and HCO_3^-.

60°C and 100°C, the deposit is a mixed calcium carbonate–magnesium hydroxide scale with the proportion of magnesium hydroxide increasing with increasing operating temperature. Analyses of scale from stills operating at 100°C and higher have shown appreciable percentages of calcium sulfate, increasing with the operating temperature. Since the rate of scale formation is not only dependent on the degree of supersaturation of the component but also on many other factors, e.g., the structure of the respective surface and often circulation patterns and retention time of the brine in the still, these rules about the dependence

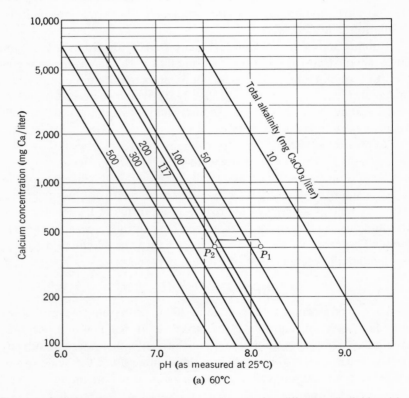

Fig. 4.5. Stability diagram for calcium carbonate for different alkalinities (a) at 60°C and (b) at 100°C. Lines were calculated for average sea water and are also valid for other water of similar composition. Each line refers to water of specified alkalinity. Region on the right of respective line represents supersaturated,

of scale composition on the operating temperature are merely approximate.

Thus, for instance, scale samples removed from thermocompression evaporators, operated at about 100°C, normally contain magnesium hydroxide almost to the complete exclusion of calcium carbonate. It is thought that calcium carbonate can stay in supersaturated solution, or at least in a fine suspension, longer than magnesium hydroxide.

In brackish waters of different character, the temperature regions for the various scale types are different, in accordance with the concentrations of the ions in the water.

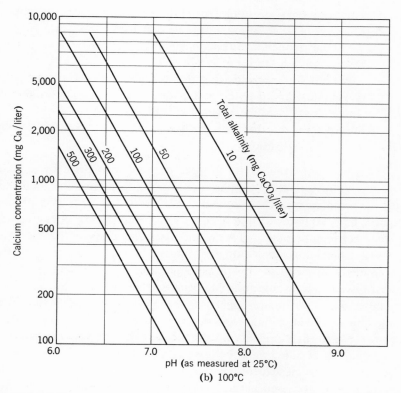

unstable solution likely to deposit calcium carbonate scale. Alkalinity is given in "mg CaCO$_3$/liter." Point P_1 represents calcium concentration and pH of average sea water. The pH difference between P_1 and P_2, 0.50, is the saturation index. From Langelier et al.

CHAPTER 4

SCALE PREVENTION

Numerous methods for the prevention of scale in *distillation processes* are known, but many of them are quite expensive. The choice of the correct method depends critically on (1) the nature of the raw water and (2) the operating conditions of the process. Less is known about scale prevention in other desalting processes.

Scale prevention methods in distillation can be grouped into several classes:

1. Design of the unit and process so that scale has no opportunity to form from the supersaturated solution before it leaves the unit.
2. Removal of scale-forming salts before distillation (softening).
3. Additions of chemicals, e.g., acids, that broaden the temperature range in which the water is stable.
4. Addition of compounds that retard scale deposition, e.g., polyphosphates, or cause the precipitation of soft, easily removable sludge, instead of scale which adheres to solid surfaces.
5. Precipitation of the scale in special, easily cleanable towers (contact stabilization).

The most important factor in scale prevention is the choice of the lowest possible operating temperature, since supersaturation increases with the temperature. Unfortunately, this method of operation usually requires large equipment. It is necessary to strike the correct economic balance between reduction of scale and operating costs on the one hand and low capital investment on the other. Much also depends on the design of the unit and the operating conditions of the process. Smooth heat transfer surfaces are desirable because they provide few nuclei for crystallization. It seems reasonable that the retention time of the water in the unit should be minimal and that there should be no pockets where it can stagnate. High circulation velocities of brine are also beneficial.

Softening of saline water before complete salt removal is sometimes required. With some exceptions, saline waters are quite hard and the cost of treatment chemicals and softening plants is considerable. Sea water contains roughly 6 kg of hardness (expressed as calcium carbonate) per cubic meter. (This is about 50 times higher than in typical surface waters treated in municipal plants.) The amount of chemicals necessary to remove this hardness is of the same order. A plant producing 5000 m^3 per day would consume several tens of tons of chemicals per day and require a large softening plant. Sometimes, however, the calcium or magnesium can be utilized in the production of bulk chemicals required in the same area as the desalted sea water, and then softening on a limited scale proves economically feasible, particularly for brackish waters of lower hardness than sea water.

Acid injection causing a decrease of pH is one of the least expensive methods of scale prevention. In Figs. 4.4 and 4.5, pH reduction corresponds to a movement to the left of the point describing the composition of the water, bringing it into a stabler region. Acid injection has practically no effect on calcium sulfate supersaturation, however, except when sulfuric acid is used. Here the supersaturation is actually increased by addition of more sulfate ions with the acid. In sea water this increase is usually negligible compared to the sulfate concentration already present.

The use of the two low-priced acids, sulfuric and hydrochloric, has one drawback: overdosing causes corrosion troubles. In large installations it is worthwhile to control the pH carefully, preferably by instantaneous-acting feedback devices. In small installations, weaker acids, e.g., citric, have been used, although they are more expensive. Instead of solutions of acids proper, it is also possible to use solutions of salts which undergo partial decomposition to acids in aqueous solutions by hydrolysis, e.g., solutions of ferric chloride at the rate of about 150 g of commercial ferric chloride hexahydrate per ton of sea water. Easy control of the dosage has been achieved by electrolytic production of ferric chloride *in situ*, directly in the sea water. In this process sacrificial iron anodes are used which dissolve in the sea water, forming first ferrous and then ferric chloride.

Many compounds have been found that *delay* scale deposition rather than prevent it entirely. The selection of a suitable product depends on the nature of the water and the operating conditions. Many of the old scale remedies have been found by experience rather than deductive reasoning. A famous, unconfirmed story tells about the discovery of starch as a scale inhibitor. Many years ago, it is said, a number of workmen repaired a very large boiler and decided to have their lunch right inside the shell. One of them forgot the potatoes which were left over, and they remained in the boiler when it was started up. It was noticed that a softer scale was formed in the next run and that the boiler could be operated for a longer period before the next cleaning.

Mixtures of corn starch, soda, and disodium phosphate have been used in water treatment for many years and found quite effective. Thus a mixture of sodium tripolyphosphate, lignin sulfonic acid derivatives, and esters of polyalkylene glycols is particularly suitable for reducing scale formation in flash distillation of sea water, provided the maximum brine temperature does not exceed 86°C. The dosage is 5 ppm in the sea water, corresponding to 10 g/m^3 of fresh water if half the water is evaporated. Sodium hexametaphosphate is also frequently used. Thus conventional water-treatment technology which has been well developed over the last hundred years often plays an important part in the planning and operation of desalting plants. Instead of polyphosphates which are *inorganic* polyelectrolytes, certain water-soluble *organic* polymers also hold promise as good scale inhibitors, e.g., polymeric carboxylic acids.

The action of many additives is probably based on colloidal phenomena. Although polyphosphates form soluble calcium complexes, the amount of this treatment agent added to the water is far too small to explain its action by this mechanism. It is believed that the beneficial action of ferric chloride is also partially due to the colloidal iron compounds formed by hydrolysis, which act as protective colloids delaying precipitation of the scale components.

Introduction of seeds of scale-forming minerals into boiling brines is quite common in many chemical industries. These seeds act as nuclei for the formation of a soft sludge within the liquid instead of the scale forming on solid surfaces. Thus in evaporators for the production of common salt, calcium sulfate scale can be prevented by maintaining calcium sulfate crystals in suspension. The seeds do not even have to be identical with the scale minerals, and many powders can act as nuclei. The *contact stabilization method* is based on the same principle. Here supersaturated brine is withdrawn from the evaporator before it has had time to deposit scale. It is pumped upward through a sand bed where precipitation occurs and is then recirculated to the evaporator. The contact bed is periodically cleaned.

The cost of scale removal can sometimes be reduced by the use of special designs of equipment, e.g., with basket-type heating elements of deeply corrugated thin Monel metal sheet. Steam condenses inside the basket and heats the sea water outside it. After a certain amount of scale has formed on the outside, the hot brine is rapidly replaced by cold water. The flat sides of the corrugations contract under this "thermal shocking," removing at least part of the brittle scale.

SELECTED LITERATURE

Chemistry of Scale Formation

Langelier, W. F., "Chemical Equilibria in Water Treatment," *J. Am. Water Works Assoc.* **38:**169 (1946).

Langelier, W. F., "Mechanism and Control of Scale Formation in Sea Water Distillation," *J. Am. Water Works Assoc.* **46:**461 (1954).

Glater, J., and Schwartz, J., "High-Temperature Solubilities of Calcium Sulfate Hemihydrate and Anhydrite in Natural Sea-Water Concentrates," *J. Chem. Eng. Data* **21:**47 (1976).

Conventional Methods of Scale Prevention

Badger, W. L., *et al.*, "Critical Review of Literature on Formation and Prevention of Scale," *Office of Saline Water Report No. 25,* U.S. Department of the Interior, Washington, D.C. 1959.

Badger, W. L., and Banchero, J. T., "Research and Development on Scale Prevention in the United States," *Proceedings of the Symposium on Saline Water Conversion, 1957,* National Academy of Sciences–National Research Council Publication No. 568, Washington, D.C., 1958, p. 44.

Neville-Jones, D., "Research and Development in Distillation and Scale Prevention in the United Kingdom," *Proceedings of the Symposium on Saline Water Conversion, 1957,* National Academy of Sciences–National Research Council Publication No. 568, Washington, D.C., 1958, p. 35.

Scale Control by "Seeding"

Chernozubov, V. B., Zaostrovskii, F. P., Shatsillo, V. G., Golub, S. I., Novikov, E. P., and Tkach, V. I., "Prevention of Scale Formation in Distillation Desalination Plants by Means of Seeding," *Proceedings of the First International Symposium on Water Desalination,* Vol. 2, Washington, D.C.,

1965, p. 539, for sale by Superintendent of Documents, Washington, D.C. 20242.

Simpson, H. C., and Hutchinson, M., "Calcium-Sulfate Scale Deposition in Sea-Water Evaporators," *Desalination* **2**:308 (1967).

Scale Control by Antiscaling Compounds

Baldwin-Lima-Hamilton Corp., "Scale Control for Saline-Water Conversion Distillation Plants," *Office of Saline Water Research and Development Report No. 186,* U.S. Department of the Interior, Washington, D.C., 1966. Deals with several proprietary compounds, containing, e.g., starch, lignin sulfonates, phosphates, and/or polymeric phosphates.

Langelier, W. F., Caldwell, D. H., and Lawrence, W. B., "Scale Control in Seawater Distillation Equipment," *Ind. Eng. Chem.* **42**:146 (1950).

Shaheen, E. I., and Dixit, S. N. S., "Scale Reduction in Saline Water Conversion," *Desalination* **13**:187 (1973). Deals with organic polymers (polyacrylates and polymethacrylates).

5

DISTILLATION METHODS

All distillation methods for saline water utilize the fact that only water and the gases dissolved in it are volatile, whereas the salts are not. If distillation is to be carried out at temperatures above 300°C, the volatility of the salts has to be taken into account. (Although such distillation methods have been suggested, they are not considered practical at the present state of our technology because of the high steam pressures and the corrosion problems encountered.) In the discussion of all practical distillation processes, we can safely assume that, upon continuous heating of salt water, merely the water evaporates, while the salts stay behind. The vapors are condensed, and thus pure liquid water if formed.

Distillation is the best-developed sea-water purification method. Most of the demineralized water produced in the world from sea water is obtained by some variant of the distillation method.

The principle of distillation is illustrated in Fig. 5.1. Sea water is boiled in the evaporator by passing hot steam through the steam chest, where the steam condenses on the inside of the tubes of the chest and is returned to the boiler. The vapors rising from the sea water are cooled in the condenser and thus converted into pure liquid water which is collected in a storage vessel. The system is vented through a pump or ejector and thus the amount and pressure of air in it can be regulated. The brine concentrate is continuously or intermittently withdrawn from the evaporator.

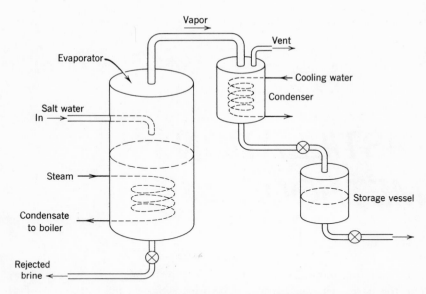

Fig. 5.1. Principle of single-stage distillation.

Instead of passing the steam through tubes surrounded by sea water, it is also possible to design the evaporator so that the sea water is on the inside of a bank of tubes, the outer surface of which is heated by steam, as shown in Fig. 5.2.

The temperature of the boiling sea water is a very important process variable, for it sets limits on the properties of the steam used. Obviously, the temperature of the condensing steam in the chest, or of any other heating medium, must be higher than the temperature of the boiling sea water. Otherwise, the large amount of heat released in the condensation could not be transmitted to the boiling water, for (in the absence of other flow processes which affect the direction of the heat flow) heat always flows from points of higher temperature to those of lower temperature, and not *vice versa*. The temperature and pressure of the *heating* steam often cannot be varied at will. It is therefore important to adapt the boiling temperature of the sea water to the properties of the heating steam by regulating the pressure in the still. This can best be understood by consideration of the vapor-pressure curves of pure and salt water.

In Fig. 5.3 the water-vapor pressures of pure water and sea water

DISTILLATION METHODS

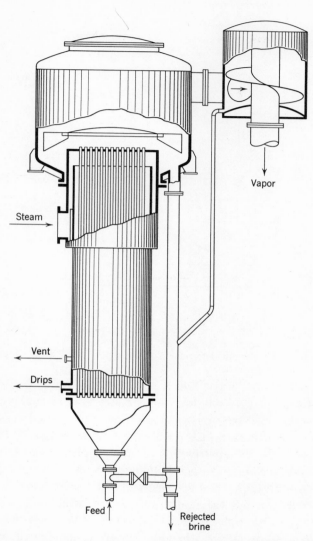

Fig. 5.2. Long-tube-vertical (LTV) evaporator. From Badger and Standiford (1958).

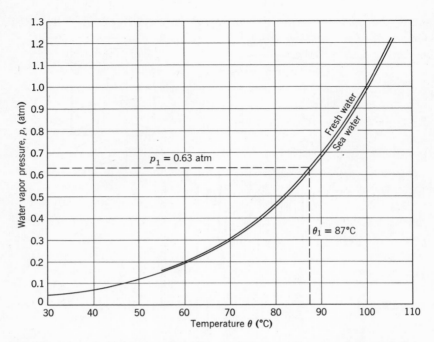

Fig. 5.3. Increase of water vapor pressure of pure water and of sea water with temperature. θ_1 is the boiling temperature at total atmospheric pressure p_1 and also the condensation temperature when the partial pressure of the water vapor equals p_1.

are plotted as functions of the temperature. These curves show the vapor pressure measured over water left in an evacuated vessel. When air is also present the *total* gas pressure rises, but the *partial* pressure of water vapor, i.e., the contribution of water molecules to the total gas pressure over the water, remains almost the same.

The water-vapor pressure rises with increasing temperature. From Fig. 5.3 we can find the temperature θ_1 at which the water-vapor pressure p_1 reaches a given level. Conversely, if heat is withdrawn from water vapor of pressure p_1 while its pressure is maintained constant by appropriate volume changes, it will condense to pure liquid water exactly at the temperature θ_1. This is also the boiling temperature of water when the prevailing atmospheric pressure is p_1, for boiling occurs when the partial pressure of the water reaches the prevailing atmospheric pressure and the rising water vapor displaces the blanket of air on the surface of the water.

We can see from Fig. 5.3 that the vapor pressure of *sea water* at a given temperature is lower and hence the boiling temperature at a given pressure higher than that of *pure water*. The vapor-pressure difference is 1.84% of the vapor pressure of pure water and the boiling-point elevation 0.46, 0.53, and 0.60°C at 75, 100, and 125°C, respectively. While these differences seem slight at first sight, they must be taken into account in multistage distillation processes, and their significance increases with the number of stages.

Steam is commonly used as a heating medium in large installations, but heating can also be effected by different means. Electric heating is convenient but expensive. In the early days of distillation technology, heating by direct flame was common, but hard-scale formation on the heating surface and overheating of the evaporator made this method obsolete. The advantages of using steam as a heat source lie (1) in the ease of controlling its temperature and (2) in the relatively large amount of latent heat which it gives up when condensing.

The *latent heat of condensation* of water at 100°C is 539 kcal/kg. This is the amount of heat given up when 1 kg steam condenses to liquid water at 100°C. The same amount of heat is necessary to convert 1 kg of water to steam at this temperature. Hence this heat is also called *heat of vaporization*. It is easy to see that this amount of heat is relatively large compared to the *specific heat* of water. It takes only 1 kcal to heat 1 kg of water by 1°C, hence the amount of heat given off when 1 kg of steam condenses at 100°C is sufficient to heat 5.4 kg of water from the freezing point to the boiling point at atmospheric pressure. Since the latent heat of melting of ice is 79.7 kcal/kg, the same amount of heat can melt almost 7 kg of ice. The heat of vaporization varies somewhat with the temperature, being 554, 539, and 522 kcal/kg water at 75, 100, and 125°C, respectively. The latent heat of vaporization of sea water is almost identical to that of fresh water, but the specific heat of normal sea water is smaller by several per cent than that of pure water, being 0.964 kcal/kg and degree Centigrade at 23°C, as compared to 0.998 for water and 0.92 for sea water concentrated to twice its original concentration.

HEAT-TRANSFER COEFFICIENTS

One of the major factors determining the size of a distillation unit is the rate of heat transfer across the heating tube between the condensing steam and the water. This rate is given by the equation

$$\mathcal{J}_Q/t = U \times A \times \Delta T$$

where the meaning and the units of the symbols in two common systems are as follows:

		Cgs system	Engineering system
$\mathcal{J}_Q$ =	amount of heat flowing from steam side to water side to tube	cal	British thermal units (Btu)
A =	area through which the heat flows	cm^2	sq ft
ΔT =	temperature difference between steam and water side of tube	°C	°F
t =	time of heat flow	sec	hr
U =	overall heat-transfer coefficient across the wall of the tube	cal cm^{-2} (°C)$^{-1}$ sec^{-1}	Btu ft^{-2} (°F)$^{-1}$ hr^{-1}

The area A is taken as the arithmetic mean between the inside and outside areas of the tube.*

The *heat-transfer coefficient* of a uniform material, e.g., copper, is simply its *thermal conductivity* k [cal cm^{-2} (°C/cm)$^{-1}$ sec^{-1} or Btu ft^{-2} (°F/inch)$^{-1}$ hr^{-1} in the cgs and engineering systems, respectively] divided by the thickness of the material. Thus the thermal conductivity of a metal is numerically equal to the heat-transfer coefficient across a sheet of thickness 1 cm (in the cgs system) or 1 inch (in the engineering system). To transform values of the thermal conductivity given in the cgs system into the engineering system, multiply by 2905.

---

*If the ratio between outer and inner diameters of the tube is less than 2, the *arithmetic* mean is sufficiently accurate. The logarithmic mean is theoretically correct.

TABLE 5.2. Heats of Combustion of Various Fuels

	Heat of combustion		Amount of fuel needed for evaporation of 1 m^3 water (100°C)
	kcal kg^{-1}	Btu lb^{-1}	
Liquid and solid			
Fuel oil	10,600	19,000	51 kg (14 gal)
Bituminous coal	6,700	12,000	81 kg
Wood (white pine, 12% water)	4,400	7,900	123 kg
Gaseous[a]			
Hydrogen	33,900	61,000	15.9 kg or 6700 ft^3
Methane	13,300	23,950	40.6 kg or 2150 ft^3
Butane	11,800	21,300	45.6 kg or 667 ft^3
Carbon monoxide	2,430	4,370	222 kg or 6750 ft^3
Heat obtained from electrical energy	860 kcal kwhr^{-1}	3,412 Btu kwhr^{-1}	625 kwhr

[a] Wet gas at 15°C and 1 atm.

heat which these fuels can yield under ideal conditions of combustion. Losses of 20% in practice are not uncommon. The figures quoted for fuel oil, bituminous coal, and pine wood are representative averages, but actual values may vary ±15% with the source of the material.

The magnitude of the fuel resources necessary for evaporation is quite impressive. For every 20 tons of water evaporated more than 1 ton of fuel oil is necessary, even if all the heat can be extracted from the oil. Moreover, consider that the salt water must be heated to the boiling point before evaporation. A considerable portion but not all, of the heat necessary for this can be supplied by heat transfer from the blowdown (rejected) brine. Considering all the losses in the boiler and evaporation plant and the steam for pumps, it is hard to design a single-effect still which used less fuel oil than 10% of the weight of the water produced. (In a *single-effect evaporator*, the vapors are simply condensed and not used for boiling additional amounts of salt water.) The fuel values of other liquid hydrocarbon fuels, e.g., kerosene, are similar to those of fuel oil, and no major savings can be expected from substituting them for fuel oil. If the still is heated by electrical heating coils, which is wasteful from an exergetic viewpoint because it degrades electrical to thermal energy, the energy requirements are about 600 times the theoretical

Fig. 5.5. Single-effect evaporator. Courtesy Maxim Division, Emhart Manufacturing Co., Hartford, Connecticut.

minimum necessary for the recovery of pure water from sea water at 50% fresh-water recovery!

Considering these high heat requirements for evaporation, the economic necessity for using the latent heat of the vapors rising from the boiling sea water for the distillation of additional amounts of salt water is obvious. The most important distillation methods—multiple effect, flash, and compression distillation—indeed utilize this principle.

Single-stage evaporation is practiced only where compact size of the plant is important and/or exhaust steam is easily available. Marine plants producing up to about 50 tons of fresh water per day manufacture fresh water by this method. Single-effect laboratory stills for the production of several tens of liters to several tons of distilled water per day are quite common. Some of the larger plants work at reduced pressure because the reduction of the boiling temperature minimizes the scale problem. A photograph of a single-effect still is shown in Fig. 5.5.

MULTIPLE-EFFECT DISTILLATION

Principle

In multiple-effect stills, the vapors from the first evaporator condense in the second, and their heat of condensation serves to boil the sea water in the latter. Therefore, the second evaporator acts as a condenser for the vapors from the first, and the task of these vapors in the second evaporator is like that of the heating steam in the first. Similarly, the third evaporator acts as a condenser for the second and so on. This principle is illustrated in Fig. 5.6. Each evaporator in such a series is called an "effect."

Obviously, the boiling temperatures and pressures in the different evaporators cannot be the same. Consider, for instance, the boiling process in the first evaporator. The *heating* steam is at a pressure of 1 atm; hence it condenses at 100°C. If the pressure in the vapor space of the first evaporator were also 1 atm, the corresponding boiling temperature for sea water would be about 100.53°C (Fig. 5.3), and the heat of condensation could not flow to the boiling sea water, since it would be at a higher temperature than the condensing steam. It is necessary to

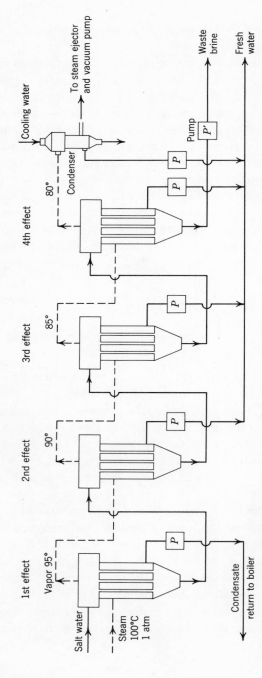

Fig. 5.6. Principle of four-effect evaporator (downflow of brine). Pressure and boiling temperature decrease from left to right.

maintain reduced pressure in the vapor space of the first evaporator, in order to make up for the difference of the boiling temperatures of pure water and salt water. This alone is not enough. To maintain reasonable heat flow rates across the pipes from the condensing steam to the boiling sea water, the temperature of the sea water must be at least several degrees lower than that of the condensing steam. If the boiling temperature of the sea water in the first evaporator is to be 95°C, the pressure in the vapor space of the first evaporator must be maintained at 0.82 atm, as seen from Fig. 5.3. At this pressure, the vapors will condense in the second evaporator at about 94.5°C, and we might wish to maintain the salt water at a boiling temperature of 90°C to provide a reasonable temperature difference across the walls of the pipes.

Since the pressure in all evaporators is less than atmospheric, pumps P must be provided to deliver the fresh water at atmospheric pressure and exhaust the steam space in the evaporators. Provision must also be made for the removal of air and other noncondensing gases from the vapor spaces of the evaporators. If these gases were allowed to accumulate, the pressure over the boiling liquid would soon rise sufficiently to stop the boiling processes. Therefore, the vapor spaces in all the evaporators could be connected to a steam ejector coupled with a vacuum pump. Instead of this arrangement, it is also possible to connect each vapor space through suitable throttles to that of the next effect where the pressure is lower, and only the last one to the steam ejector and vacuum pump. The latter method is possible where the amount of air is small. It causes less loss of steam with the exhausted air than the first. Pump P' withdraws the waste brine from the last evaporator and restores it to atmospheric pressure.

Figure 5.6 does not show various other improvements in the heat economy of the process which are possible and which are worthwhile in large plants. The incoming sea water is heated by serving as coolant in the final condenser, and the sensible heat of the fresh water and the waste brine is also used to preheat the sea water. The condensate from the steam ejector which maintains the partial vacuum is combined with the mass of fresh water. Instead of parallel flow of sea water and vapor (from left to right in Fig. 5.6), countercurrent flow can be used; i.e., the salt water enters the last evaporator first. In this case, additional pumps must be provided to move the salt water between the evaporators.

The principle of multiple-effect evaporation makes it possible to produce more than 1 ton of distilled water per ton of heating steam. In fact, the amount of fresh water produced per unit amount of heating steam increases almost in direct proportion with the number of stages. Experience in many installations has shown that the number of tons of fresh water produced per ton of heating steam in one-, two-, three-, four-, and five-effect evaporators is 0.9, 1.75, 2.5, 3.2, and 4.0, respectively.

It is obvious that the larger the number of effects, the lower the heating steam requirements per ton of fresh water produced. On the other hand, increasing the number of effects causes higher investment costs. As in many other industrial processes, the relation between the operating expenses and fixed charges determines the optimum size of the plant.

Optimum Number of Effects

In principle, many effects are possible. The condensation temperature of the heating steam always has to be slightly higher than that of the boiling salt water, but since the boiling-point elevation (i.e., the temperature difference between boiling sea water and fresh water, respectively, at equal pressure; see Fig. 5.3) is relatively small this consideration alone does not seriously limit the number of effects. Thus in a still operating between temperatures of 100°C and 27°C in the first and last effects, respectively, concentrating normal sea water twofold, ΔT_B contributes only 0.7°C on the average per effect. Hence an ideal multiple-effect plant could incorporate over 100 stages! In practice, however, it is necessary to maintain a temperature difference considerably larger than 0.7°C across the heat-transfer surfaces so that the heat flows at a reasonable rate.

The major factors determining the cost of plants of given production rate of pure water are (1) the fixed costs, namely, interest on invested capital and taxes and depreciation and repairs, all of which are approximately proportional to the number of effects, and (2) the daily cost of heating steam, which is roughly inversely proportional to the number of effects used. Thus the cost C per unit of water produced is expressed by an equation of the form

$$C = An + \frac{S}{n}$$

where A is the fixed costs for each effect per unit of water produced, S is the cost of steam per unit of water produced in a single-effect evaporator, and n is the number of effects.

By differentiating C with respect to n and setting the derivative zero, it is found that the optimum number of stages is

$$n_{opt} = \sqrt{S/A}$$

Thus the whole number nearest to $\sqrt{S/A}$ will represent the optimum number of stages. This calculation is only approximate since it neglects the costs of the condenser, steam ejector, and vacuum pumps, and also routine labor costs. Because these plants are highly automated, the total labor cost per plant does not increase very much with the size of the plant. In other words, the proportion of labor in the unit cost of water decreases appreciably with increasing size of the plant.

A fair number of large land-based multiple-effect evaporators have been built in various regions. In these installations, the number of effects is usually six or less. The low ratio between fuel costs and investment is due partly to the low price of fuel in some of these locations and partly to the high cost of the construction materials. The heating tubes must be made from copper, nickel, brass, admiralty metal, or other alloys which are good conductors of heat and corrosion resistant. All these materials are relatively expensive. Because of its superior corrosion resistance, titanium has also been used in spite of its relatively high cost and has replaced the copper alloys in some plants.

Alternatively, small amounts of sulfuric acid can be added to the feed in order to reduce the pH and prevent precipitation of scale. This method increases the corrosivity of the feed and is therefore not ideal for continuous use. It is quite effective against calcium carbonate and magnesium hydroxide, but not against calcium sulfate scale, which appears only at higher temperatures. On the other hand, the addition of polyphosphates to sea water, while more expensive, can also inhibit the precipitation of calcium sulfate and does not increase the corrosive properties of raw sea water. It works best at temperatures below 82°C (180°F).

FLASH DISTILLATION

Principle

In the distillation processes discussed in the previous sections, heating of the salt water and boiling take place in the same vessel. Some scale can form in sea water when it is merely heated and no appreciable evaporation has yet taken place, but most of the scale forms during evaporation proper. In the flash-evaporation process, sea water is first heated in tubes and then made to evaporate in chambers in which a pressure lower than in the heating tubes prevails. Since the vapor simply flashes off the warm liquid, the resulting precipitates form primarily in the liquid.

At first sight, the flash process might seem rather inefficient because hot water cools off considerably when only a small fraction has been evaporated. For instance, the evaporation of only 7.1% of a given amount of water initially at 100°C causes it to cool to 60°C. On the other hand, the advantages of minimizing the scale problem and the simplicity of the design of flash evaporators have made this method a close competitor of regular multiple-effect evaporation, and in large plants frequently even more economic than the latter.

For instance, most of the sea-water stills in the Persian Gulf region are flash units. Kuwait, which had the largest production of desalted water in 1974 [about 240,000 tons per day (65 million gallons per day) installed and almost 65,000 tons per day under construction] uses mostly flash-distillation units, and the largest single distillation unit in the world being put through trial runs in 1976 was a flash-distillation plant in Hong Kong with a production capacity of almost 200,000 tons per day.

The principle of the flash-distillation process is shown in Fig. 5.7, which also illustrates the advantage of performing the flash operation in several stages. Salt water enters a bundle of tubes located in the vapor

DISTILLATION METHODS

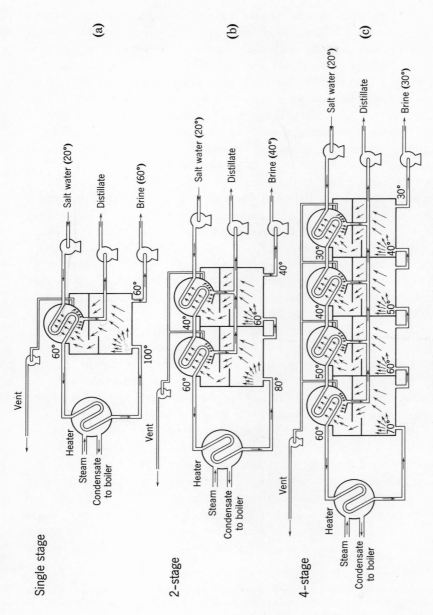

Fig. 5.7. Principle of multiple-stage flash evaporation.

space of the flash chamber where it is preheated. It then passes into a heater consisting of a bundle of tubes heated externally by steam where the temperature of the salt water is raised to 100°C, but since the pressure is kept higher than 1 atm, no boiling occurs. The *hot* sea water then enters a flash chamber kept under reduced pressure. Part of the water evaporates, the vapors condensing on the tubes carrying the incoming *cold* sea water to the heater. Distillate and brine are restored to atmospheric pressure by pumps. The tubes across which heat exchange* between condensing steam and incoming sea water occurs play an important role in the economy of the process.

The temperature of the brine leaving the flash chamber depends on the amount of vapor flashed off the chamber. If it is desired to recover 7.1% of the salt water as distillate, we must let it cool to 60°C in the flash chamber. Hence the water-vapor pressure prevailing in the flash chamber will be that corresponding to salt water at 60°C (Fig. 5.3). The vapor reaching the cold sea-water tubes will also condense at about 60°C. Hence the highest temperature to which the sea water can be preheated in this chamber is somewhat below 60°C and the remaining heating to 100°C must be done in the heater.

It is seen that the heat requirements of a single-stage flash evaporator are about equal to those of a regular single-effect evaporator of the same production rate, since the amount of heat used for evaporation in the single flash chamber is equal to the heat of condensation of the heating steam. In other words, the weight of condensate produced in the flash chamber is roughly equal to the weight of heating steam condensed.

It is possible, however, to produce more distillate per unit of heating steam if the flashing is carried out in more than one stage. In Fig. 5.7b, a two-stage flash unit producing the same amount of distillate as the single-stage unit is schematically shown. The temperature of the evaporating salt water in the two flash evaporators is 60 and 40°C, respectively, and the incoming brine is heated by 20°C in each unit. The amount of water evaporated in each unit is half of the corresponding amount in the single-stage unit. The brine is discharged at 40°C instead of 60°C as before.

---

*The common term *heat-exchange* is used with the understanding that heat flows from the higher to the lower temperature, and that no give-and-take "exchange" as discussed in Chapter 8 ("Ion Exchange") is involved here.

Hence less heat is lost with the discharged brine. Consequently, less heat is introduced into the heater in which the temperature of the salt water is raised from 60 to 80°C, not 100°C as in the single-stage unit. Therefore, the amount of heating steam required is only half of the amount in the single-stage unit.

Finally, in the four-stage unit (Fig. 5.7c) the flashing process is carried out in four stages, reducing the temperature of the evaporating brine and increasing that of the incoming salt water by 10°C each. The brine is discharged at 30°C and the amount of heating steam required is only one-fourth of the amount in the single-stage unit because the heater raises the temperature of the salt water by only 10°C.

All the temperatures shown in Fig. 5.7 have been calculated with rough approximation, since the purpose of the figure is merely to illustrate that the heat economy improves with the number of stages. In practice, heat transfer is never as perfect as assumed, hence salt water in the tubes leaves all stages colder than the condensing distillate. In fact, the overall heat-transfer coefficient is generally lower than in multiple-effect evaporators because the salt water in the flash evaporator tubes does not boil. In practice, the heat consumption of a flash evaporator does not decrease as rapidly with the number of stages as in the simplified diagram of Fig. 5.7 or as in multiple-effect stills. On the other hand, it is possible to add more stages to a flash evaporator without using separate shells for each stage and the costly connections, that are necessary in regular multiple-effect evaporators.

Heat Economy

Practical plants incorporate a number of additional features to improve the heat economy and reduce the formation of scale and corrosion.

The feed is treated with about 5 ppm of "Hagevap LP" [a product composed of polyphosphates and lignin sulfonate (Calgon Corp., Pittsburgh, Penn.)], or a similar product, to promote the formation of a soft sludge of calcium carbonate instead of adherent scale. Corrosion is minimized by passing the feed water through a deaerator before it enters the heater. In the deaerator a large portion of the air and carbon dioxide contained in the water is released and sludge precipitates.

Air and other noncondensing gases are continuously withdrawn through a vent system in order to maintain the pressures in the different stages constant at their prescribed levels.

Since a major portion of the investment cost is represented by the tube material, it is customary to consider the economy of flash evaporation in terms of the size of this heat-transfer area. Increasing the number of stages at constant performance ratio, i.e., ratio of weight of total distillate produced to heating steam used, causes a decrease in the required heat-exchange area. Twenty stages are common for a *performance ratio** of 8. Comparing installations containing a large number of stages and having identical daily output, it can be shown that the performance ratio is proportional to the total heat-transfer area in the installation. This again means a roughly inverse relationship between operating costs and investment.

The heat-transfer tubes can be arranged vertically or horizontally. Either of these arrangements has its particular advantages, and the latter is preferable for units of large capacity. Figure 5.8 shows the chamber construction principle of a 20-stage horizontal-tube flash evaporator, arranged in two tiers. Each tier is split by a longitudinal vertical partition wall, and there are five flash chambers on each side of this wall in each tier. It is seen that the design of these stills is very compact.

In principle, it is possible to operate flash evaporators over a wide range of pressures and temperatures, but in practice the highest temperature of the salt water which occurs at the exit from the heater is not much higher than 100°C. Therefore, reduced pressure must be continuously maintained in the flash chambers. The reasons for the relatively low operating temperature of the flash process are (1) the desirability of using low-pressure steam as a heating medium and (2) freedom from scale deposits in the salt-water tubes. These considerations also apply to the multiple-effect evaporators described in the previous section, e.g., long-tube vertical evaporators.

Heating steam of pressure between 1.5 and 1.0 atm is considered "low-pressure steam" in industrial usage. The corresponding condensa-

---

*The performance ratio is defined as the number of pounds of distillate produced per 1000 Btu heat input in the heater (Fig. 5.7). Since the heat of evaporation of water is approximately 1000 Btu/lb, the performance ratio is approximately equal to the amount (by weight) of distillate produced per amount (by weight) of steam used in the heater.

DISTILLATION METHODS

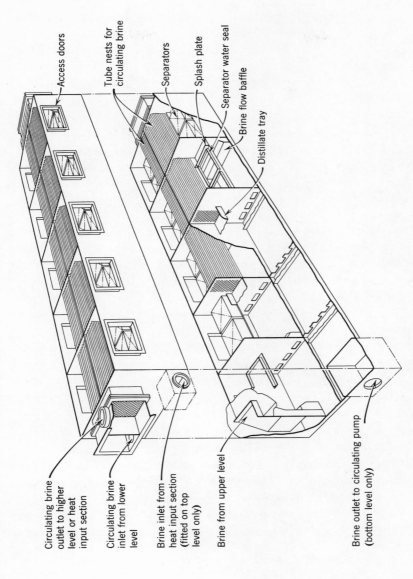

Fig. 5.8. Horizontal flash evaporator. From Frankel (1960).

Fig. 5.9. Multiple-stage horizontal-tube flash-distillation plant. Production capacity 4000 tons per day (about 1 mgd). Photo courtesy Israel Desalination Engineering (Zarchin process), Ltd., Tel Aviv, Israel.

tion temperatures are 112°C and 100°C, respectively, and are quite sufficient for operation of the heater in a flash evaporator or the first effect in a multiple-effect still. Such steam is often available at low price from chemical plants. It is also possible to couple steam-driven electrical power plants using fossil or nuclear fuel with salt-water distillation plants. The use of steam from a low-temperature nuclear reactor has also been suggested and a detailed economic analysis has been carried out. Low-temperature reactors can be built at considerably less expense than high-pressure power reactors.

Figure 5.9 shows a horizontal-tube multistage flash plant of capacity 4000 tons per day at Eilat, Israel.

Vapor-Reheat Process

The heat-exchange surfaces in conventional flash evaporators must necessarily be large, for the temperature difference between the condensing vapors and the sea water in the tubes is small. Because of the high cost of these tubes, attempts have been made to modify the flash process in order to reduce the size of the heat-exchange surfaces required.

In the *vapor-reheat flash-evaporation process,* also called *direct-contact distillation process* shown schematically in Fig. 5.10, the vapors in the flash chambers are condensed on a spray or film of cold distilled water. The contact between the vapors and the distilled water film is direct and no metallic surfaces separate them, but the salt water and distilled water in the flash chambers are not allowed to mix. The temperature of the distilled water rises as a result of the condensation of vapor. The distilled water stream is pumped from each stage to the adjacent stage operating at higher temperature and pressure. Its volume increases continuously as condensation of the vapor takes place and reaches a maximum in the first flash chamber. This hot distillate is now passed through a countercurrent liquid-liquid heat exchanger ("product water cooler") in which it first transfers the heat gained in the passage through the flash chambers to an immiscible oil, and then the latter, in turn, transfers the heat to the incoming cold sea water in another exchanger ("sea-water heater"). These heat-transfer processes, too, take place by direct contact and without metallic surfaces. In the heat exchangers the light oil rises through the water, which flows downward.

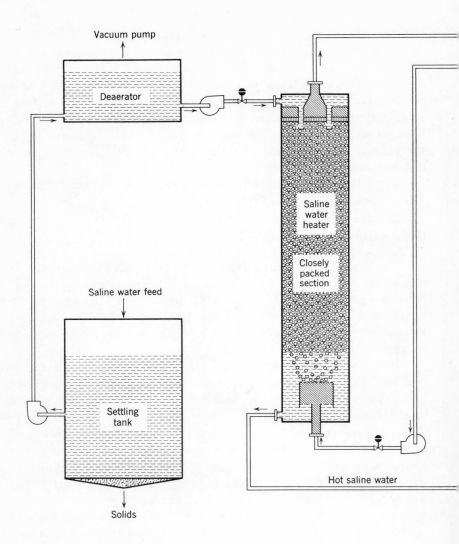

Fig. 5.10. Principle of vapor-reheat flash-evaporation process. Saline water is treated by settling and deaeration, then heated by countercurrent flow of hot oil in direct-contact liquid–liquid heat exchanger. Water then passes down distillation column. Vapor is

DISTILLATION METHODS

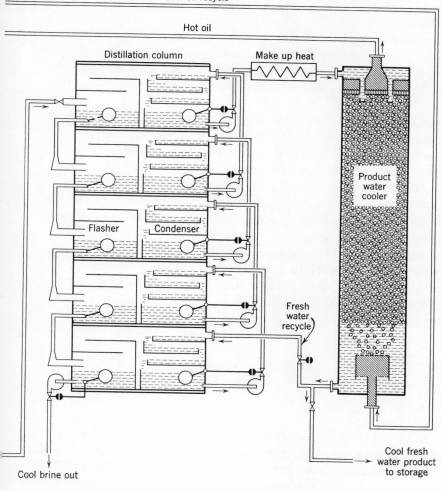

collected by stream of recycled fresh water. Hot fresh water transfers heat to oil in second rect-contact heat exchanger. Heat necessary to maintain continuous process is introduced into fresh-water stream. Courtesy FMC Corporation, Santa Clara, California.

An amount of distilled water equal to the flash steam condensed in all of the stages is withdrawn as product. The balance is recycled to act again as a condensing medium for the vapor in the flash chambers. To provide the energy necessary for the process and to make up for losses, heat is added to the system in the section marked "make-up heat." Conventional heat exchangers may be used instead of the immiscible oil type units.

It is seen that this process is very similar to regular flash distillation except that distilled water acts as a direct coolant for the vapors instead of the tubes carrying cold salt water. The heat accumulated by the distilled water is transferred to the sea water in a separate heat-transfer device which need not have metallic heat-transfer surfaces.

Different variants of this method have been suggested. Some use solid rock beds for heat storage and cool fresh water rather than oil as a vapor-condensing medium.

The vapor-reheat process is an interesting method because it applies unconventional approaches to heat-transfer problems in distillation. It is not known, however, if it will ultimately prove more economical than the conventional flash evaporation method described in the previous paragraph.

VAPOR-COMPRESSION DISTILLATION

Principle

While multiple-effect and flash-evaporation units use an external supply of heating steam as the primary heat source, vapor compression distillation, which is often shortened to "compression distillation," uses literally its own steam, after it has been compressed, as a heat source. In this method, it is possible to obtain high power economy, but it is necessary to provide mechanical energy by means of mechanical compressors (or another form of "available energy" by other devices, e.g., steam-ejector compressors). Although the desalting process is different here from the idealized process described in Fig. 3.1, it is noteworthy that in both cases the driving power which makes desalting possible is contributed by a compressor rather than by heat sources as in the other distillation processes discussed in this chapter.

DISTILLATION METHODS

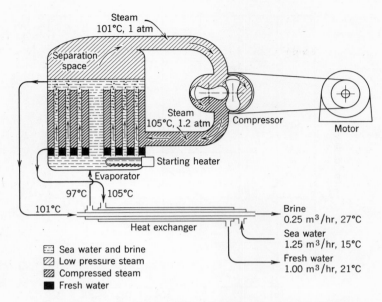

Fig. 5.11. Principle of compression still. From Latham (1946).

As shown schematically in Fig. 5.11, sea water, preheated in a tubular heat exchanger by the outgoing streams of brine and fresh water, boils in the tubes of the still. The vapors are compressed, and led back to the still to condense outside the tubes, thus providing the heat necessary for the boiling process. The noncondensing gases are withdrawn from the steam-condensation space by a suitable vent pump or ejector.

The heart of the installation is the compressor. If the vapors were not compressed, they could not condense on the tubes carrying the boiling sea water because the condensation temperature of pure vapor at a given pressure is less than the boiling temperature of the salt water at this pressure. For instance, if the vapor pressure is 1 atm, water vapor condenses at 100°C, but doubly concentrated sea water boils at about 101°C. In order to make the vapors condense at 101°C, it is necessary to compress them to at least 1.03 atm.

It is seen that in this process the energy is supplied by a compressor, often a direct-displacement unit driven by an engine or electrical motor. It is also possible to use steam ejectors as vapor-compressor devices, but in small installations the energy used is mechanical, except for the small

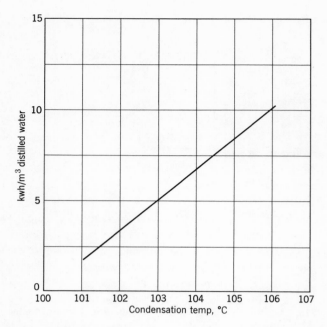

Fig. 5.12. Compression work requirements for compression distillation. Doubly concentrated sea water boils at 101.05°C. Vapors are compressed from 1 atm to higher pressure and condense accordingly at a higher temperature. Graph shows compression work required per cubic meter of water produced as a function of this condensation temperature. Work amounts are for adiabatic compressor of 100% efficiency.

amount of heat necessary to speed up the start of the operation. When calculating power requirements it is well to keep in mind that the temperature of vapors rises when they are compressed. In fact, if the water vapors were compressed in a perfectly insulated container ("adiabatic compression"), the rise in temperature would be so great that it would more than make up for the raising of the condensation temperature by the increased pressure. In other words, the vapor would now be too hot to condense.* It is the presence of the cooler tubes in which the salt water boils at a constant temperature that causes the condensation of the compressed vapor in the vapor chest of the compression still.

The work W_c necessary to compress 1 ton of superheated steam of

*Conversely, if nonsaturated water vapor is expanded adiabatically, the cooling effect is sufficient to make it condense in spite of the decreased pressure, as, for instance, in the Wilson cloud chamber used in the study of high-energy radiation.

temperature 101°C and pressure 1 atm to r atm is given by

$$W_c = 24.0 \times \frac{r^2 - 1}{r} \text{ kwhr/m}^3 \text{ fresh water}$$

The derivation of this equation is given in Appendix 3A, which also contains a more complex equation applicable to other temperatures and pressures of the superheated steam. In Fig. 5.12, W_c is plotted against the condensation temperature corresponding to r, i.e., the temperature of the fresh water condensing in the steam chest. This temperature must be larger than the boiling temperature of the twice concentrated brine within the tubes, about 101°C.

It is seen that the work requirement increases linearly with the condensation temperature. To minimize the power requirements it is desirable to work at the lowest possible condensation temperatures, but then the rate of heat transfer across the tubes is low and many tubes are needed. Obviously, it is necessary to weigh operating costs against investment and thus determine the optimum design conditions for the plant. This has been done and a number of basic designs were evolved, some of which are described in the following paragraphs.

Small Units

Although the basic idea of the compression still is over 100 years old, accelerated research and development leading to very significant advances started just before World War II. Small units to provide drinking water for ships were developed in compact size. These units produce between 0.5 and 10 m³/day and are constructed according to the principle shown in Fig. 5.11 or with the sea water boiling outside and the steam condensing inside the heating element. Power consumptions for electrically driven small installations are between 40 and 100 kwhr/m³ of fresh water produced from sea water. The smaller the unit, the larger the power consumption per unit product. This is largely due to the relatively high losses of heat through the insulation in the small-capacity units. Even in the larger units of this type, only about half the electrical power is used for the compressor, the other half being consumed by the sea-water heater to make up for all the heat losses.

To make the stills entirely independent of electrical power supplies,

Fig. 5.13. Small compression still. This still is driven by an internal-combustion engine and its output rated at 0.5 m³ fresh water per hour. Unit without trailer weighs 1.3 tons, with trailer 2.7 tons. The trailer-mounted unit can be lifted by helicopter. Courtesy Badger Manufacturing Company, Cambridge, Massachusetts.

units with compressors driven by internal-combustion engines were developed. A photograph of such a unit rated at about 0.5m³ of fresh water per hour is shown in Fig. 5.13. This unit can be powered by a diesel engine. The heat of the exhaust gases from the engine driving the compressor is utilized for auxiliary heating of the sea water. This heat is more than sufficient to perform the same function as the electric heaters in electrically driven units.

The total power consumption in the electrically driven units of this size is at best 25–30 times and the power for the compressor above 12–15 times the theoretical minimum to produce fresh water in an ideally reversible process. In the units driven by gasoline or diesel motors, 1 kg of fuel produces up to 200 kg of pure water. To obtain this fuel efficiency in a multiple-effect evaporator, we would have to use at least 15 stages, but in the latter case less expensive fuels than gasoline or diesel oil can be used. Many kinds of small stills have been developed. Because of the compact design and their independence of electrical power sources, the engine-driven small compression stills have found wide use in the United States Navy during World War II.

The evolution of the Zarchin direct-freezing process (Chapter 7) has necessitated the development of special compressors that can move large volumes of low-pressure vapor. Such compressors, which are large fans with flexible blades, are also useful for compression–distillation units in which no ice is formed but which operate at considerably lower temperatures than 100°C, when scale formation presents a less serious problem. It is seen from Fig. 5.3 that the pressure in such stills is less than atmospheric. The temperature difference across the heat-transfer surface is only a few degrees Centigrade. At the low temperatures, this corresponds to a small pressure difference. For instance, the water vapor pressure difference between 103°C and 100°C is 0.11 atm, but the corresponding difference between 53°C and 50°C is only 0.019 atm (Fig. 5.3). Thus when operating at low temperatures only a small pressure difference prevails between the evaporating and steam-condensing spaces. This small difference, in turn, makes it possible to use relatively simple centrifugal or axial blowers. In other words, compression can be carried out by a device similar to a ventilating blower.

A serious problem in the operation of vapor-compression stills at elevated temperatures is the rapid formation of scale on the seawater side of the heat-transfer surfaces. Since the temperature difference across these surfaces is small in the first place, any decrease in the heat-transfer coefficient by scale formation is a serious impediment to the process. To maintain constant production, the compressor has to work harder as scale keeps accumulating. When used with ordinary sea water, the early units had to be shut down for cleaning after every few hundred hours of operation, when the compressed steam pressure rose by 50–100%.

To obviate these difficulties, and to minimize corrosion, commercial "Aquaport" units [Israel Desalination Engineering (Zarchin Process), Ltd., Tel Aviv] were designed to operate below 45°C. In these units, saline water is sprayed on the *outside* of a bundle of horizontal heat-transfer tubes, forming a thin, continuous film, while compressed vapor condenses on the *inside* of the tubes. The locations of brine and condensing vapor are therefore opposite to those in the unit of Fig. 5.11.

Solutions to the scale problem primarily applicable to small stills were found in scraping devices, which clean the surfaces continuously, and in acid injection. When designing large units both the scale problem and the reduction of power requirements to the bare minimum must be kept in mind. Such considerations are also incorporated in the design of small low-temperature units, which operate at subatmospheric pressure and often incorporate flexible-blade compressors, utilizing centrifugal forces for compression, as described in the following section.

Centrifugal Compression Stills

The fundamental innovation in these stills as developed by K. C. D. Hickman is a rapidly rotating heat-exchange surface over which the evaporating brine spreads as a thin film that causes a high heat-transfer coefficient, even at low temperatures. It is possible to operate these stills efficiently at temperatures as low as 50°C, at which scale formation is not a serious problem.

The principle of the centrifugal compression still is shown in Fig. 5.14. Salt-water feed enters the partially evacuated space bounded by two rotating cones. Both points of application of the water to the inner surface are near the axis of rotation. The water spreads by centrifugal action and evaporates partially while the vapors are compressed by a blower which rotates at a higher speed than the double cone. The compressed vapor condenses at the outer side of the cone, providing the heat for the evaporation of the salt water on the inside. Concentrated brine is continuously withdrawn from the rotor periphery by a scoop tube. In order to maintain sufficiently low pressure at all times, it is necessary to degas the feed water entering the still and also to continuously remove some of the gases in the steam space. This bleed-off is adjusted to remove both air and, inevitably, some water vapor, and in the steady state it maintains at a comfortably low concentration the amount of air which enters the apparatus through imperfect degassing of the feed and mechanical leaks.

The drawing of Fig. 5.14 is merely schematic and leaves out many details. To improve the yield, practical units have heat transfer to the incoming feed from the outgoing brine and distillate. It is possible to replace the conical rotors by flat disks which spread water at least as well as the conical variety. One or many disks or cones can be mounted horizontally on a vertical shaft.

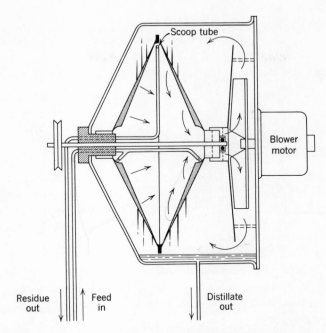

Fig. 5.14. Principle of centrifugal compression still. From Hickman (1958).

Past experience with these stills was derived primarily from a number of small experimental units, but also from a sizable research unit ("Hickman No. 5 still") which was operated for several months. This unit contained eight copper rotors each 2.4 m in diameter mounted on a vertical shaft rotating about 400 times per minute. Operating conditions were varied, and the optimum production rate with this still was 60 m³/day produced from 200 m³ sea water. The boiling temperature was about 55°C and the condensing temperature 52°C. In spite of the low temperatures, the heat-transfer coefficient across the rotor wall was 2100 Btu ft^{-2} (°F)$^{-1}$ hr^{-1} or more than three times higher than in conventional heat-exchanger tubes working at about 100°C. The energy consumption was 17 kwhr/m³, of which 9.5 kwhr was used for the blower and the rest for the rotor, pumps, and other accessories.

Still higher heat-transfer coefficients were obtained with smaller units. Thus in a laboratory unit incorporating various improvements, heat-transfer coefficients up to 4500 Btu ft^{-2} (°F)$^{-1}$ hr^{-1} were observed. In fact, from many points of view, small units seem to offer the best prospects for these rotary stills. They are subjected to smaller centrifugal stresses and might, in the future, prove quite valuable for domestic use.

COMBINATION OF DISTILLATION WITH POWER PRODUCTION

It is sometimes economical to combine salt-water purification with power production, or to combine different desalting methods so that the cost of the water is less than if each had been performed separately. Such plants are often called "dual-purpose" plants.

When a new community is established in an area remote from electrical power and fresh-water sources (or if both power and water production in a given area are to be increased), and if salt water is locally available, it is possible to use the exhaust steam from the back-pressure turbine-generators of the power plant as a heat source for distillation. For instance, a six-effect distillation plant in Aruba is combined with a power plant.

In conventional power plants, high-pressure steam drives turbines and is then condensed in surface condensers cooled by cold fresh water or sea water, whichever is readily available. If, instead of the latter operation, the exhaust steam is led into a flash distillation plant and condenses in the heater section there, the efficiency of power production is reduced because the exhaust steam condenses at a higher temperature and pressure. Hence the pressure drop across the turbine and the power production per kilogram of steam are less. For example, if a 50,000 kw power plant driven by superheated steam of 90 atm and 500°C discharges the steam at 1.7 atm into a flash-distillation plant, instead of condensing it by cold salt water to a pressure of 0.04 atm, the steam consumption of the power plant rises by 56% from about 4.1 to 6.4 kg/kwhr of which 5.3 kg can go to the flash-distillation plant, the remainder being used for feed-water preheating in the power plant. How much fresh water can be produced from this amount of steam? In conventional flash-distillation plants, each ton of heating steam can easily produce 5 tons of fresh water. By increasing the number of stages, even better yields are obtainable. Design studies have been made for plants with yields as high as 14 tons of fresh water per ton of heating steam. Assuming a more conservative tenfold yield, the amount of fresh water generated would be 53 liters/kwhr. Thus in a plant working at average 65% load, 0.83 m^3 fresh water could be produced daily per kilowatt installed capacity. In other words, each daily cubic meter of water corresponds to approximately 1.20 kw installed capacity. Compare this to the Aruba six-effect combination plant designed for 29 kwhr/m^3 water, which is a yield similar to that of the hypothetical flash-distillation plant.

It is of interest to ask if the amount of water obtained as a by-product of power production would be sufficient for the people using the power. In the highly industrialized society of the United States, the installed power is about 0.75 kw per person. If the same power capacity is installed in a hypothetical new desert community, $0.75 \times 0.83 = 0.62$ ton of water can potentially be produced per person and day. This is somewhat less than twice the average domestic consumption of a member of such a society, but corresponds to only *one-sixth* of the amount of water used in the United States for all purposes except irrigation. Hence the amount of by-product would not even be sufficient to cover the needs of the industries using the power, unless (1) those industries which use relatively large amounts of water, e.g., paper and pulp, were excluded, (2) a serious attempt were made to substitute salt water for fresh water wherever possible, and (3) fresh water were reused to the fullest extent. There is no hope at present of obtaining sufficient by-product water for a fully developed agriculture producing grain, rice, or similar bulk crops.

The cost of fresh water produced in large dual-purpose plants would in general be lower than in distillation plants which have to purchase or generate steam independent of power production.

At first sight, one wonders if an advantage can be gained by actually reducing the *power* production of a plant by using the distillation plant rather than the cold sea water as a condenser. Indeed, no power advantage would be gained if instead of a distillation plant, another power plant (operating between an intermediate temperature and the final condenser fed by cold sea water) were interposed between the primary superheated steam and the final condenser. It should be kept in mind, however, that the partially exhausted steam (pressure 1.7 atm in the example given here) is needed here for its *caloric* rather than its *power-producing ("exergetic")* value, because the power-producing potential of the steam fed into a distillation plant is not realized; the power necessary to split salt water into fresh water and brine (discussed in Chapter 3 and Appendix 1A) is much less than the equivalent of the flow of heat entering the distillation plant in the form of driving steam.

Steam from nuclear power plants may be used as the heating medium for multiple-effect stills or for multiple-stage flash-distillation plants. Such dual-purpose plants are called *nuclear desalination plants*, although no nuclear process is involved in the desalination section of the plant. Where *geothermal* steam is available, it may be used for the same purpose. Heat for power production and/or desalination can also be extracted from geothermal brines, but these hot saline liquids are often too corrosive for direct use, because of very high salinity and the presence of corrosive gases, e.g., hydrogen sulfide. Therefore, *binary fluid cycles* are being considered, in which the hot brine from the geothermal well is first used as a heating medium for a second, less corrosive working fluid for power production in turbines and/or as a heating medium for a desalination plant. In 1976, the use of geothermal heat for desalination was in the development stage, but no production plants were in operation.

In addition to combination processes involving power production, a number of internal combinations of desalting processes have been designed and equipment is commercially available. For instance, compression stills can be coupled to multiple-effect or flash stills. Multieffect evaporation can be combined with multistage flash evaporation. Such combinations often prove somewhat more economical than plants based on one particular method only.

SOLAR EVAPORATION

Since the cost of heat plays a decisive role in distillation processes, it seems attractive to harness the heat of the sun for this purpose. Many such devices have been invented and tested over a period of almost 100 years. A solar still covering about 4000 m^2 and providing fresh water in an arid area was built in Chile in 1872 and operated for many years.

DISTILLATION METHODS

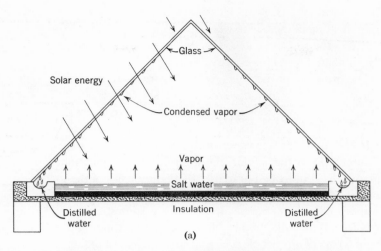

Fig. 5.15. (a) Principle of solar still. Radiation heats salt water in black trough. Vapor condenses on sloping glass surfaces and runs off into distilled water troughs. (b) Simple solar still. Pan area about 3 m². The performance of this still was studied by author at Weizmann Institute of Science, Rehovoth, Israel, 1948–1950. Slanting surfaces are glass. Back plates were coated with aluminum paint. Photo courtesy F. Schleissner, Rehovoth, Israel.

During World War II, small plastic solar stills to provide fresh water for liferafts floating in the ocean were developed. There has been progress in this field in the last three decades, due to the general increase of interest in the utilization of solar energy.

The principle of the solar still is illustrated in Fig. 5.15a. Salt water contained in a black pan covered with a sloping glass roof is heated by the sun. Water vapor rises to the glass where it condenses, forming a film which runs off into a collecting trough and from there to storage.

In this type of still, water does not boil, but vaporizes slowly, and the vapors reach the cooler glass surface by convection. The rate of evaporation is controlled primarily by the intensity of the incoming solar radiation. If there is no radiation, the salt water in the pan rapidly cools, its vapor pressure is reduced, and the evaporation process virtually comes to a standstill. We are often inclined to think about evaporation rates in terms of air turbulence or wind velocity. In reality, however, the most important factor controlling solar evaporation is the amount of solar heat received per unit surface of the evaporating liquid per unit time.

This amount can be increased by concentrating onto the surface additional solar radiation, by means of lenses, mirrors, and other focusing devices. We cannot, however, concentrate more energy into the liquid than the collecting surface intercepts, and it often proves more expensive to cover an area with suitable lenses and mirrors than with additional pans for evaporation. Hence the simple device shown in Fig. 5.15 has proved quite popular and has been studied by many investigators.

Although the energy for solar evaporation is free of charge, the installation costs are considerable, since only a few liters per day can be produced per square meter of pan area in the still shown in Fig. 5.15b. Therefore, it is very important to design stills for maximum utilization of the incident energy. To estimate the possible yields and the unavoidable losses, it is necessary to know the amount of solar radiation received at the site of the proposed distillation plant.

Incidence of Solar Radiation

The amount of solar energy received per unit horizontal surface is measured by means of recording pyrheliometers. These instruments consist essentially of black and white metal strips placed under a glass dome in a vacuum. The black strips absorb more radiation than the white ones and hence their temperature is higher. The temperature difference between the black and white strips is recorded and indicates the total radiation. These measurements are often available from the respective meteorological services and provide the basis for predictions of solar-still yields in new locations.

The total yearly amount of solar energy received on a horizontal surface in different places is listed in Table 5.3. These figures include all solar radiation: (1) the visible, which comprises somewhat less than 40%; (2) the ultraviolet, which is less than 5% and varies considerably with the humidity of the air because water vapor absorbs ultraviolet radiation; and (3) the infrared radiation, which, although invisible, contributes 55–60% of the energy. The figures are given in several units, including the depth of the water layer, which could be evaporated per day if the solar-still efficiency were 50%, i.e., 50% of the total incident radiation as measured by the pyrheliometer were utilized for evaporation.

We can intercept more solar radiation by *rotating* the receiving surface continuously so that the radiation is always normal to the surface. For instance, at geographical latitude 35°, we could thus intercept 58% more radiation. This relatively small gain could hardly justify the expense for the turning mechanism in simple solar distillation devices; however, stationary stills tilted toward the horizontal plane at an angle equal to the latitude intercept more solar energy than horizontal devices. At 35° latitude the average gain is about 16% of the energy received on a horizontal surface. Such a still has been developed by the solar energy use pioneer, Dr. Maria Telkes, and is shown in Fig. 5.16.

It is interesting to compare the amount of heat received per unit evaporating surface in a simple solar still to the heat transferred across heating tubes, in a regular still. Suppose the heat-transfer coefficient in the latter is 500 Btu ft^{-2} (°F)$^{-1}$ hr^{-1} or 120,000 Btu is transferred per square foot and day when the temperature difference across the tubes is 10°F (5.5°C). This is 120 times more than the solar heat used for evaporation in Algeria, assuming that 50% of the incoming radiation is utilized. This comparison shows that solar stills must be made considerably cheaper, per unit heat-transfer area, than conventional stills in order to be competitive in initial investment. The operating cost of solar

TABLE 5.3. Average Solar Radiation Received on Horizontal Surface at Different Places[a]

Location	Btu ft^{-2} day^{-1}		cal cm^{-2} day^{-1}	Depth of evaporation, cm day^{-1} (50% efficiency)[b]	U.S. gal ft^{-2} and day (50% efficiency)
	Variation	Yearly average			
El Paso, Texas	1200–2730	2030	551	0.485	0.119
Tamamassit, Algeria	1460–2400	2000	542	0.477	0.117
Poona, India	1430–2690	1980	537	0.473	0.116
Phoenix, Arizona	1061–2710	1925	522	0.460	0.113
Messina, South Africa	1340–2320	1875	508	0.447	0.110
Nice, France	571–2525	1500	407	0.358	0.088
Salt Lake City, Utah	443–2192	1442	392	0.345	0.085
Boston, Massachusetts	502–1938	1240	336	0.296	0.073
London, England	181–1740	882	239	0.210	0.052
Antarctica	0–735				

[a] From Löf (1958). Variation refers to average daily radiation in months of lowest and highest evaporation, respectively. All other figures are yearly averages.
[b] Heat of evaporation taken as 570 kcal/liter corresponding to an average water temperature of 25°C.

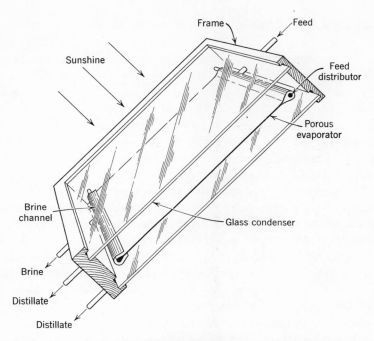

Fig. 5.16. Flat tilted solar still. Schematic cross-section. From Telkes (1958).

stills is, of course, much less because the heat is free. Although the amount of water produced in solar stills per unit surface area seems small from this comparison, it is well to keep in mind that large areas can be covered with solar stills. For instance, 1 km^2 of still area in Algeria, which might correspond to an area of 1.3 km^2 for the total installation, will produce 1.7 million m^3 of fresh water per year if the caloric efficiency is 50%. In other words, a layer of water 1.7 m deep would evaporate yearly from the pan area. If solar evaporation were to provide irrigation water and if the amount of irrigation water necessary were equivalent to 60 cm rainfall per year, about 1 km^2 of evaporator pan area would have to be created for each 3 km^2 of irrigated land.

In these estimates, 50% efficiency has been assumed. This is actually a hopeful extrapolation into the future. Efficiencies measured in present experimental installations are lower, ranging between 20% and 45%. Calculations show that an annual average efficiency of about 50% presents an upper limit for operation of roof-type solar stills even under very favorable climatic conditions.

Efficiency and Losses in Roof-Type Stills

The question arises to what extent solar heat losses can be minimized. These losses depend (1) on the meteorological conditions, e.g., irradiation, air temperature, and wind velocity, and (2) on the design of the still. The first cannot be changed, whereas the most important variables in the second are the nature of the materials used, shape and orientation of the still, and depth of the evaporating layer, which affects the daily temperature variation of the salt water, which in turn has a decisive influence on the losses.

To calculate the solar-still efficiency expected from the meteorological variables for a given location, it would be necessary to make hour-to-hour computations of the brine temperature and the heat losses to the surroundings. Allowance must also be made for heat storage in the brine because the brine temperatures at the beginning and end of the hour are usually different. Certain approximative methods for similar calculations of evaporation from *open* pans have been worked out. Some such calculations were also carried out for *covered* solar stills, but these calculations are generally tedious. Instead, rough estimates of the losses are usually made by using *monthly averages* of the meteorological variables. This method is best suited for solar stills with "deep" brine basins, containing brine layers of about 30-cm depth, because here the daily variation of brine temperature is less than in shallow-basin stills which warm up rapidly in the morning and cool off quickly at night. In the shallow basins, distillation during the day hours is faster than from deep layers, but it comes practically to a standstill at night. On the other hand, deep layers have more heat storage capacity. They warm up more slowly than shallow layers and reach their maximum temperature later, but distillation continues during at least part of the night hours. The deeper the layer, the less variation of the brine temperatue during a 24-hr cycle and the longer the time lag between the daily maximum of irradiation, which usually occurs around noontime, and that of the brine temperature.

Figure 5.17 shows the results of this type of analysis for a hypothetical large solar still in the San Diego, California, area, latitude N 33°. In these calculations, it was assumed that sea water is concentrated to twice its original concentration, the sea water supply undergoes heat exchange to distillate temperature, the brine absorbs 95% of the radia-

DISTILLATION METHODS

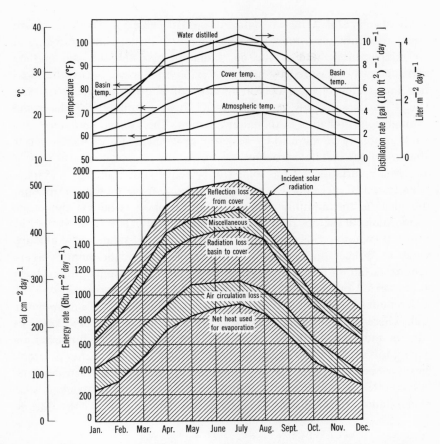

Fig. 5.17. Estimated performance of a deep-basin solar still in San Diego, California. From Löf (1960).

tion transmitted through the cover, and the still is placed directly on the ground without any thermal insulation under the pans. The last assumption derived from the consideration that the soil will act as its own heat insulator after an initial limited warm-up period. [Radiation losses are assumed to increase with the fourth power of the absolute temperature (Stefan's law).]

It is of interest to compare these predictions with actual loss measurements of a glass-covered still of this kind, carried out in October 1959 in Port Orange, Florida (latidude N 29°), as part of a systematic study of solar stills. These results are listed in Table 5.4. The losses were computed from experimental measurements of brine, vapor, cover, and air temperatures, and of the net radiation between the still and its surroundings, measured by means of a radiometer. Although the location and meteorological conditions are not quite the same as in San Diego, the results are very similar to those predicted for the same average solar radiation at the latter location. The "unaccounted losses" could probably be reduced by design improvement, for example, better shading of the distillate troughs.

Many variations of the solar-still principle shown in Fig. 5.15 have been tested. Experience has shown that the decisive factor determining the yield is the monthly average solar radiation, whereas the other meteorological factors play only a secondary role. This is illustrated in Fig. 5.18, where yields are plotted against radiation. While a plot of daily yields vs. solar radiation shows considerable scatter, the monthly average yields fall on a straight line, although the experiments plotted here were carried out in two widely separated locations and with stills of different design. The still used in Port Orange, Florida, was a deep-basin still, whereas the Australian still was a shallow-basin type developed in Algeria and made of fibrocement. It would be unjustified, however, to expect this efficiency from each solar still. Occasionally, higher efficiencies have been recorded, for example, in stills in Mildura, Australia, and in the author's experiments in Rehovoth, Israel, with a still backed by a plane stationary aluminum-paint covered reflector (Fig. 5.15b). More-

TABLE 5.4. Energy Balance for a Solar Still[a]

Energy	Percentage of solar radiation
Evaporation	32
Heat loss to ground	2
Solar radiation reflected by still	12
Solar radiation absorbed by cover and condensate film	10
Radiation from water in basin	25
Internal convection (air-circulation loss)	7
Unaccounted losses, probably reevaporation of distillate	12
	100

[a]From Bloemer et al. (1960). Location: Port Orange, Florida. Time: Oct. 7–9, 1959 (72 hr), average solar radiation 380 cal cm^{-2} day^{-1} (1400 Btu ft^{-2} day^{-1}).

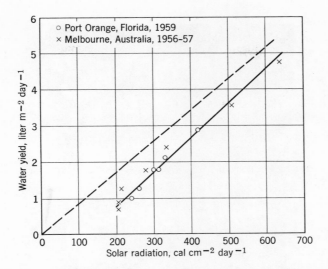

Fig. 5.18. Yields of solar stills. Figures on coordinate axes refer to monthly averages. Dotted line shows yield of hypothetical still of 50% efficiency. From Bloemer et al. (1960); Wilson (1958).

over, improvements of the materials should lead to better efficiencies in the future.

The hypothetical "50% efficiency line" has also been drawn in Fig. 5.18. It represents the yields from a still utilizing one-half of the incident radiation (as measured by a pyrheliometer) for distillation. We can see that all measured points fall below this line, but the percentage utilization of the solar energy increases with increasing energy input. In fact, if the heat absorbed by the brine would not change with the hour of the day and were continuously as high as during the peak period at noontime, efficiencies of 65–75% might easily be achieved as shown by calculations and experiments with continuous electrical rather than solar heating.

Yield Improvement

It is seen from Fig. 5.17 and Table 5.4 that the major energy loss is the long-wave radiation form brine to cover. There seems to be little that can be done to reduce this loss except a judicious choice of brine depth to minimize the daily averages of this loss in accordance with the

Fig. 5.19. Solar stills at Daytona Beach test site, Port Orange, Florida. Photo courtesy Office of Saline Water, U.S. Department of the Interior.

meteorological conditions. Heat losses to the ground are small, provided that the ground area is large and stays dry to retain its good insulating properties. The convection loss is related to the distillation rate, and changes of circulation rates will affect both in a similar manner (Appendix 4A). On the other hand, reflection of radiation and absorption in the glass can be reduced by the choice of better and/or specially treated transparent cover materials. It is possible to treat glass to reduce its reflectivity, and light absorption within the cover can be reduced by the use of thin films instead of glass. In fact, many plastic-covered stills have been designed, built, and studied, and one still installation covering 200 m² was being systematically tested together with the deep-basin stills of similar area at Port Orange, Florida, under sponsorship of the Office of Saline Water (now part of the Office of Water Research and Technology), United States Department of the Interior (Fig. 5.19). Other development activities in this field are being carried out by private industry.

The use of plastic films instead of glass seems economically very attractive at first sight. The plastic stills are kept inflated by slight excess pressure inside the still. Many plastic stills, however, have shown appreciably lower yields than glass stills because of the occasional development of leaks and the dropwise condensation of water vapor on the underside of the water-repellent film. These droplets grow to considerable size before running off into the distillate channels and sometimes drop back into the brine. Treating agents have been developed which make the films less water repellent and prevent deterioration of certain films under irradiation. It is interesting that dropwise steam condensation in fuel-powered distillation is beneficial, whereas in solar distillation the opposite is true. This fact illustrates the diversity of the methods necessary to meet the problems of modern salt-water purification development.

Other Types of Solar Stills

Because of the relatively low yields of solar stills per unit area when compared to other distillation methods, attempts have been made to utilize the incident solar energy for distillation several times before discharging the waste heat. In the simple, roof-type still, the heat of condensation of the vapor on the glass cover is wasted, being carried away by the air and by radiation to the atmosphere. Multiple-effect solar stills have been patented, but data or realistic tests are scant.

With devices more elaborate than the simple solar stills described, much higher water temperatures can be attained. By using parabolic or cylindrical mirrors to concentrate solar energy, steam can be produced to drive turbines and conventional desalting plants. The collection and concentration of solar radiation and the attainment of very high temperatures are fascinating subjects, but they are not treated here, since these devices have generally proved too expensive for salt-water purification. In solar stills with radiation-concentrating devices, the heating and distillation stages are frequently separated, whereas in the simple stills shown in Figs. 5.15 and 5.16 radiation absorption, evaporation, and condensation take place in the same device.

Cost of Stills

Experience has shown that the initial cost of solar stills is considerably higher than that of equipment for other desalting methods.

On the other hand, the simple design of roof-type solar stills made of metal or fibrocement and glass makes these units suitable for the production of limited amounts of distilled water in remote locations which have access to saline but not to fresh water. Thus substantial solar stills have been used with varying success on the island of Symi (Greece) and in the mining area of Cooper Peddy in Australia. They are, in general, not suitable for use by outposts of the armed forces because of their considerable vulnerability. On the other hand, the combination of greenhouse-type agriculture with solar distillation has been suggested and might prove to be of interest under special conditions.

SELECTED LITERATURE

Multiple-Effect Tube and Multistage Flash Evaporators

Standiford, F. C., "Evaporation," *Chemical Engineer's Handbook,* R. H. Perry and C. H. Chilton, eds., 5th ed., McGraw-Hill, New York, 1973, Chapter 11, pp. 27–38.

Badger, W. L., and Standiford, F. C., "Large-Scale Multiple-Effect Sea-Water Evaporation Plants," *Proceedings of the Symposium on Saline Water Conversion 1957*, National Academy of Sciences–National Research Council, Publ. 568, Washington, D.C., 1958, p. 103.

Eshaya, A. M., and Dodge, B. F., "Application of Forced-Circulation and Drop-Wise Condensation Techniques to Large-Scale Distillation of Sea Water," *Proceedings of the Symposium Saline Water Conversion 1957*, National Academy of Sciences–National Research Council, Publ. 568, Washington, D.C., 1958, p. 73.

Silver, R. S., "A Review of Distillation Processes for Fresh-Water Production from the Sea," *Fresh Water from the Sea, Dechema Monographie*, Vol. 47, Verlag Chemie, Weinheim-Bergstrasse, West Germany, 1962.

Bairamov, R. B., Ataev, Ya. A., and Tairov, B. D., "Analysis of Thermal Schemes for Evaporative Desalination Plants," *Izv. AN Turkm. SSR Ser. Fiz.-Tekh. Khim. Geol. Nauk*, No. 4, 1975, Soviet Union, translated in *Int. Chem. Eng.* **16** (3):387 (1976).

Van der Mast, V. C., Read, S. M., and Bromley, L. A., "Boiling of Natural Sea Water in Falling-Film Evaporators," *Desalination* **18**:71 (1976).

Design of Flash-Distillation Plants

Frankel, A., "Flash Evaporators for the Distillation of Sea Water," *Proc. Inst. Mech. Engrs. (London)* **174**:312 (1960).

Mulford, S. F., "Low Temperature Flash Distillation of Sea Water," *Proceedings of the Symposium on Saline Water Conversion 1957*, National Academy of Sciences–National Research Council Publ. 568, Washington, D.C., 1958, p. 91.

Distillation Plants in Kuwait

Reside, J., and Al-Adsani, A. M. S., "The World's Largest Desalting Complex—A Report of Twenty Years' Experience," *J. Natl. Water Supply Improvement Assoc.* **1**:1 (1974).

Shuhaibar, Y. K., "Near East Desalting," *Desalination* **17**:69 (1976).

Hong Kong Desalting Plant

NWSIA (National Water Supply Improvement Association) *Newsletter* (October 1976), P.O. Box 8300, Fountain Valley, Calif. 92708.

Drake, F. A., "Desalting in Hong Kong—The First Phase," *Desalination* **18**:1 (1976).

Use of Surfactants to Increase Heat Transfer

Sephton, H. H., *Proceedings of the Fifth International Symposium on Fresh Water from the Sea,* Alghero, Sardinia, 1976, for sale by A. A. and A. E. Delyannis, Tsaldari St. 34, Athens-Amaroussion, Greece.

Vapor-Reheat Process

Othmer, D. F., Benenanti, R. F., and Goulandris, G. C., "Vapor Reheat Flash Evaporation without Metallic Surfaces," *Chem. Eng. Progr.* **57(1)**:47 (January 1961).

Woodward, T., "Vapor Reheat Distillation," Chapter 4 in *Principles of Desalination,* K. S. Spiegler, ed., Academic Press, New York, 1966.

Compression Distillation

Latham, A., "Compression Distillation," *Mech. Eng.* **68(3)**:221 (1946).

"The Aquaport—Fresh Water from the Sea," Israel Desalination Engineering, P.O. Box 18041, Tel Aviv, Israel.

Hickman, K. C. D., "Centrifugal Boiler Compression Still," *Ind. Eng. Chem.* **49**:786 (1957).

Hickman, K. C. D., "Development of the Centrifugal Compression Still," *Proceedings of the Symposium on Saline Water Conversion, 1957,* National Academy of Sciences–National Research Council Publication No. 568, Washington, D.C., 1958, p. 51.

Dual-Purpose Plants

Aschner, F. S., Yiftah, S., Glueckstern, P., Frank, G., and Lavie, A., "Feasibility of Nuclear Reactors for Sea-Water Distillation," 1964–1967, International Atomic Energy Agency, Vienna, Austria, Contract 252/RB, Microfiche IAEA-252-F, 1968.

d'Orival, M., *Water Desalting and Nuclear Energy,* Verlag Karl Thiemig KG, Muenchen 90, West Germany, 1967.

Vilentchuk, I., "Sea Water Conversion Requires Nation-Wide Planning," *Aqua* **1**:53 (1961).

Use of Geothermal Heat

El Ramly, N. A., Peterson, R. E., and Seo, K. K., "Geothermal Wells in Imperial Valley, California: Desalting Potentials, Historical Development and a Selected Bibliography," *J. Natl. Water Supply Improvement Assoc.* **1:**31 (1974).

Laird, A. D. K., "Water from Geothermal Resources," *Geothermal Energy,* P. Kruger and C. Otte, eds., Stanford University Press, Stanford, Calif., 1973.

Solar Evaporation

Bloch, R. M., Farkas, L., and Spiegler, K. S., "Solar Evaporation of Salt Brines," *Ind. Eng. Chem.* **43:**1544 (1951). Open pans for salt recovery.

Bloemer, J. W., Collins, R. A., and Eibling, J. A., "Field Evaluation of Solar Sea Water Stills", *Advances in Chemistry Series,* No. 27, American Chemical Society, Washington, D.C., 1960, p. 166.

Bloemer, J. W., Eibling, J. A., Irwin, J. R., and Löf, G. O. G., "Solar Distillation—A Review of Battelle Experience," *Proceedings First International Symposium on Water Desalination,* Vol. 2, Office of Saline Water, U.S. Department of the Interior, Washington, D.C., 1965, p. 609.

Howe, E. D., "Solar Distillation," *Solar Energy Research,* F. Daniels and J. A. Duffie, eds., University of Wisconsin Press, Madison, 1955.

Gomella, C., "Traitement et utilisation des eaux saumâtres dans les régions arides," *Salinity Problems in the Arid Zones,* (Proc. Teheran Symp.), Documents and Publications Service UNESCO, Paris, 1960, p. 331.

Löf, G. O. G., Demineralization of Saline Water with Solar Energy," *Research and Development Progress Report No. 4,* Office of Saline Water, Department of the Interior, Washington, D.C., 1954.

Löf, G. O. G., "Design and Cost Factors of Large Basin-Type Solar Stills," *Proceedings of the Symposium on Saline Water Conversion 1957,* National Academy of Sciences–National Research Council Publ. 568, Washington, D.C., 1958, p. 157.

Löf, G. O. G., "Design and Operating Principles in Solar Distillation Basins," *Advances in Chemistry Series* No. 27, American Chemical Society, Washington, D.C., 1960, p. 156.

Löf, G. O. G., "Solar Distillation," *J. Am. Water Works Assoc.* **52:**578 (1960).

Telkes, M., "Fresh Water from Sea Water by Solar Distillation," *Ind. Eng. Chem.* **45:**1108 (1953).

Telkes, M., "Solar Still Theory and New Research," *Proceedings of the Symposium on Saline Water Conversion 1957,* National Academy of Sciences–National Research Council Publ. 568, Washington, D.C., 1958 p. 137.

Wilson, B. W., "Solar Distillation Research and Its Application in Australia," *Proceedings of the Symposium on Saline Water Conversion 1957,* National Academy of Sciences–National Research Council Publ. 568, Washington, D.C., 1958, p. 123.

Ronchaine, J. F. M., "Combinasion de la climatisation et du dessalement," *Bull. Recherches Agronomiques de Gembloux, Belgique* (hors série, 1971). Combination of greenhouse-type agriculture with solar distillation.

Some General Journals on Solar Energy

Solar Energy, published quarterly by Pergamon Press, Headington Hill Hall, Oxford OX 3 OBW, U.K., for the International Solar Energy Society, P.O. Box 52, Parkville, Victoria, Australia 3052.

Applied Solar Energy (English translation of *Geliotekhnika*, Allerton Press, 105 Fifth Avenue, New York 10011.

Solar Energy Digest, P.O. Box 17776, San Diego, Calif. 92117.

6

ELECTRODIALYSIS

PRINCIPLE

In all *distillation* methods, water vapor is removed from salt water and condensed away from the main body of the liquid. Because most water desalinated in 1976 was *sea water,* and because distillation was considered the best method for this purpose, more than four-fifths of the desalting capacity of the world was based on variants of distillation. As the desalination of less brackish water moves to the foreground, membrane methods which are more economical for waters considerably less salty than sea water are growing in importance. For instance, the large Wellton-Mohawk saline irrigation return flow to the Colorado River in Arizona will be treated by membrane methods to prevent excessive salination of the lower Colorado River which serves part of northern Mexico. Moreover, membrane methods can also be used for the desalination of sea water. While the *feasibility* of this process has been demonstrated for both electrodialysis and reverse osmosis (Chapter 9), the *economic prospects* for large-scale sea-water treatment by reverse osmosis alone look quite promising, and it is possible that this membrane method might emerge as a serious competitor of distillation methods, especially in locations where "waste heat" is not inexpensively available.

Electrodialysis was historically the first membrane method developed for water desalination and is still an important method in use today. In *electrodialysis,* the ions forming the salt are pulled out of the salt water by electrical forces and concentrated in separate compartments.

The higher the salinity of the raw water, the more electrical power is needed for this process. Hence this process is applied primarily in the treatment of moderately brackish waters containing several thousands parts per million of dissolved salts, a salinity too high for most domestic and industrial uses, yet about one order of magnitude less than that of average sea water.

The principle of electrodialysis is illustrated in Fig. 6.1. Electrodialysis units consist of a number of narrow compartments through which saline water is pumped. These compartments are separated by alternating kinds of special membranes, which are permeable to positive ions (cations) or negative ions (anions), respectively. The terminal compartments are bounded by electrodes for passing direct current through the whole stack. In saline waters the salts exist as separate positive and negative ions, as indicated in Fig. 6.1a, which shows the stack before passage of the current.

When the electrodes are connected to a direct-current source, ion travel begins and hence an electrical current passes through the unit, as shown schematically in Fig. 6.1b, which focuses attention to a group of compartments well within the stack. In the center compartment, positive ions travel from right to left, negative ones in the opposite direction, and both kinds leave the compartment through the membranes. If each membrane were permeable to both kinds of ions, no concentration change would result because for each positive ion leaving the center compartment into the left compartment another positive ion would enter from the right. Similarly, the amount of negative ions in the center compartment would also not change. The anion-permeable membrane on the right, however, does not admit cations from the right compartment to replenish the center compartment and the cation-permeable membrane on the left similarly acts as a barrier for the negative ions in the left compartment. As a result, the salt concentration decreases in the center compartment and increases in the neighboring compartments, as shown in Fig. 6.1c.

In practical electrodialysis, in installations which contain from ten to hundreds of compartments between one pair of electrodes, the passage of electrical current thus creates fresh water ("diluate") and brine ("concentrate") in neighboring cells. In other words, half of the cells carry partly desalted water and half carry brine. The solutions in the

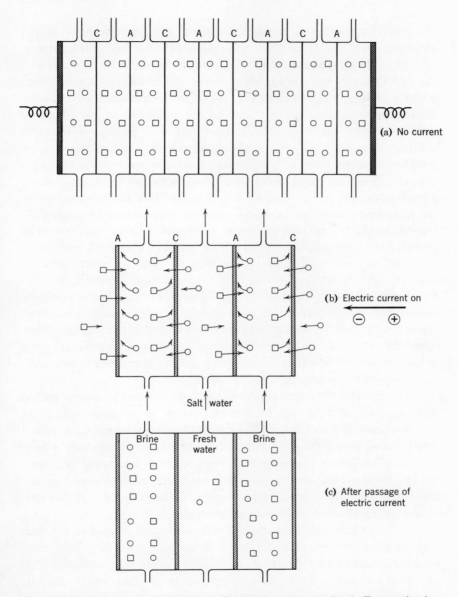

Fig. 6.1. Principle of electrodialysis. ○, Positive ion (e.g., sodium); □, negative ion (e.g., chloride); C and A, cation- and anion-permeable membranes, respectively. Ion migration under action of electrical current causes salt depletion in alternate compartments and salt enrichment in adjacent ones.

electrode compartments are contaminated with the products of the electrode reactions which occur as a result of the passage of the current.

The selective permeability of the membranes can be explained by the fact that they are ion-exchange materials. In such materials either positive or negative ions can easily move within the solid, whereas the ions of opposite charge are bound to the solid. When placed into a solution, cation exchangers exchange freely with positive ions in the solution, but practically no negative ions from a dilute solution or positive ions in excess of those exchanged can enter the solid. As a result, a cation-exchange membrane acts as a barrier for negative ions. Electrical current passes through it by the motion of positive ions which are picked up on one face and released at the other, but negative ions are barred. Similarly, an anion-exchange membrane conducts by motion of negative ions, since its interior is out of bounds for positive ions.

The ion-selective membranes are the most sensitive part of the unit. Like the heat-exchange surfaces in distillation, they determine its cost; they can foul by deposition of scale which must be dissolved or removed mechanically, and even the most durable ones require occasional replacement. Hence much effort has been invested in producing suitable membrane materials and many different types are on the market, ranging in thickness from less than 0.1 to 1 mm and in appearance from parchment paper to synthetic upholstery coverings.

The principle of electrodialysis has been applied to plants with a daily capacity of thousands of tons of fresh water, to household units designed for less than 100 liters per day, and to many intermediate sizes. Theories of transport processes in ion-exchange membranes which existed even before the advent of electrodialysis and were primarily applied to biomembranes have undergone much extension and refinement as a result of the availability of and the interest in the synthetic ion-exchange membranes used in electrodialysis.

One can carry out electrodialysis in *stationary* solutions, but then large salt-depleted layers are formed near the membrane faces where the solutions become very dilute; these layers of dilute solutions offer high electrical resistance. It is therefore customary to pump water continuously through the unit. In practical units, plastic spacer materials are placed between the membranes which force the water to take a tortuous path and thus promote good mixing within each compartment.

It is possible to shift the desired amount of salt from the diluate compartments to the adjacent concentrate compartments in a single pass; however, this mode of operation usually necessitates high currents and very large units. Moreover, there are large differences in the current densities in the unit. At the top, the low concentration of the diluate causes high electrical resistance and hence low current.* At the bottom, the concentration of the liquids in the two compartments is the same and the overall resistance lower, hence higher current densities. The whole process is easier to control, and more economical, if the desalting process is not carried out in a single pass, but either in a series of membrane stacks, mounted separately, or in a single frame. In small units, the partially treated diluate and brine are sometimes collected in separate tanks and recirculated through the membrane stack several times until the desired degree of desalting is achieved.

The power requirements for the process are entirely electrical. They can be grouped into two parts: (1) direct current for ion shifting and (2) power for pumps, which are usually driven by alternating current.

POWER REQUIREMENTS

In electrodialysis, as in all other desalting processes, a minimum amount of "available energy" is required to split the raw water into fresh water and brine. At first, it seems as if this requirement could be made arbitrarily small, by passing a very low current at an arbitrarily low voltage through the stack. This is impossible, however, because a _minimum voltage_ is necessary to pass a current in the desired direction. When an electrodialysis unit is switched off at any stage of the desalting process, it is observed that an electrical voltage exists between the electrodes, which always opposes the externally applied voltage. In other words, the unit acts now like a battery. If current is withdrawn from it, it is found that the ions move opposite to the directions in the electrodialysis process, namely, from the brine compartments into the fresh-water compartments, until the concentrations are equal again. (This process takes place in addition to slow salt diffusion through the

---

*The decreased resistance of the concentrate (because of its high salt concentration), usually does not make up for the increase of resistance in the diluate compartment.

membranes, which also tends to equalize the concentrations.) For instance, in a stack of 100 membrane pairs, containing brine at a concentration of 5000 ppm and fresh water of 500 ppm, this countervoltage is about 12 volts. To carry out electrodialysis, the applied voltage must be at least somewhat larger than this countervoltage, which expresses the minimum power required for desalting in electrical terms. In practice, the opposing voltage is often even larger, because of polarization phenomena, which are discussed in a later section. The opposing voltage is termed *polarization voltage*.

The polarization voltage observed in practice is usually caused by electrode reactions as well as concentration differences and certain undersirable side reactions at the membrance surfaces, all of which cause the establishment of voltages opposing the passage of current. Moreover, if the total voltage applied were only of the order of 0.1–0.2 volt per membrane pair contained in the stack, as in this example, desalting would proceed at a very slow rate and extremely large units would be required. To reduce the capital charges, it proves more economical to operate at higher voltages, of the order of 1 volt per membrane pair, although this entails a loss of electrical power because the faster motion of the ions causes relatively more conversion of electrical energy into heat.

In more quantitative terms, the amount of salt shifted per hour and unit of membrane stack cross-section and the hourly yield of fresh water are proportional to the current density, i amp cm^{-2}, whereas the electrical power consumption, in watts, is proportional to Ri^2. R is the resistance of the stack in ohms.* The exergy dissipation per kilogram of salt shifted is therefore roughly proportional to the current. In other words, for a given raw water, the faster the unit operates, the higher the d.c. (direct-current) power consumption per unit of fresh water production rate. Hence it is convenient to define the following ratio:

$$\text{Power index} = \frac{\text{electrical power (d.c.) per unit fresh water rate}}{\text{production rate divided by total membrane area}}$$

The production rate is reduced to equal membrane area in order to make comparison between units of different size possible. Since both

---

*The exact expression is $Ri^2 + V_p i$, where V_p is the polarization voltage and $V_p i$ the power to overcome this voltage. This power consumption refers only to direct current (d.c.).

numerator and denominator are roughly proportional to the current, the power index should be independent of the applied voltage, current, or rate of production. Certain deviations from this constancy result because of the changes of the stack resistance with temperature and polarization, but the power index is still a simple criterion for comparison of different units. If a manufacturer of electrodialysis units states that his installation consumes less kwhr per m^3 of fresh water, this does not necessarily mean that his unit is superior, for by decreasing the production rate of other units a lower power consumption can also be achieved. Therefore, the power consumptions of the different units have to be reduced to the same production rate per unit membrane area and this is exactly what the power index does. Of course, this comparison must also be done for equal raw water and product salinity, for in electrodialysis, as opposed to distillation, the power needs vary greatly with the salt concentration.

To compare the performance of units for different degrees of salinity reduction, it is useful to calculate the power index per 1000 ppm salt (as NaCl) removed. This provides a measure of the amount of d.c. electrical energy for unit rate of salt removal effected by unit membrane surface. This comparison, however, is meaningful only if the desalting ranges to be compared are not too different, because the more dilute the solutions, the higher the electrical power demands for the shifting of a given quantity of salt per unit time. (This is true since the voltage necessary to maintain the ion flow is higher when the solution is more dilute.) For example, it takes considerably more electrical power to reduce the salt concentration from 1300 to 300 ppm than from 4000 to 3000 ppm, although in both cases the concentration reduction is 1000 ppm.

In practical units, the power index varies between about 10 and 50 watt hr^2 ft^2 gal^{-2}, but when reduced to the same salt removal [i.e., to 1000 ppm salt (as NaCl)] these figures reduce to about 10 in most instances. The average electrical current densities used increase with the salinity of the water. They are usually of the order of 100 amp per square meter (10 milliamp per square centimeter) of membrane surface. It is of interest to compare this value with electrical current densities used in conventional electrochemical devices, such as electrolytic cells for the production of chlorine and alkali, or storage batteries used in automobiles. In these devices, all of which contain more concentrated solutions, the electrical current densities are one to two orders of magnitude higher. This fact emphasizes that the membrane areas needed for a given ion-shifting rate are much larger than the electrode-metal surfaces in conventional devices producing an equivalent rate of an electrochemical reaction. Hence relatively large membrane areas are needed for electrodialysis and the cost (per unit area) of the membranes is very important for the economy of this process.

The power index refers only to direct-current requirements. Additional power is required mainly for pumping, and also for instruments and controls. The amount of pumping power depends appreciably on the design of the units, in particular the kind of spacer used. Many spacers which force the water streams into highly tortuous paths promote good mixing and reduce polarization, but they also cause high power requirements for pumping. In most plants actually in operation now, the total power requirements are from about one and one-half to double the amount of d.c. power required for electrodialysis alone.

OPTIMIZING PLANT PERFORMANCE

When determining the most economical operating conditions of an electrodialysis plant, we are faced with the following dilemma: is it desirable to (1) acquire few membrane stacks and operate them at maximum production rate or (2) acquire many and operate them in parallel at lower than maximum rate? In the first case, investment and interest will be low, but power costs high; in the second, the reverse is true.

The most economical size of the plant and power consumption for a given fresh-water production rate (m³/day) and raw-water and fresh-water salinity can be determined by the following approximate analysis.

The total cost of unit amount of fresh water produced, d, is the sum of three terms:

$$d = ai + (b/i) + c$$

where, a, b, and c are constants and i is the current density.

The first term represents charges which are proportional to the production rate, i.e., chiefly the electrical power consumption, which is roughly proportional to the current density. This refers both to d.c. electrodialysis power and to pump power because the latter is proportional to the pumping rate, which in turn is proportional to the production rate. All charges quoted refer to unit amount of product.

The second term of the equation represents the charges which are proportional to the number of stacks, i.e., interest amortization, insurance, and similar charges. These are inversely proportional to the production rate because the larger the daily production in a given plant, the less the investment per cubic meter. The third term, independent of current density, includes costs of chemical treatment and other charges which are largely independent of the number of stacks used.

To find the optimum current density, the equation is differentiated with respect to i and the derivative set zero. This yields the optimal current density:

$$i_{opt} = \sqrt{(b/a)}$$

Substituting this value in the original equation, it is found that when the most economical operation is achieved, the first two terms are equal. In other words, the number of units must be chosen so that the annual power costs are equal to the annual sum of interest, amortization, insurance, and similar fixed charges.*

The current densities used in the desalting of brackish waters containing up to 5000 ppm dissolved solids generally lie between 6 and 20 milliamp cm^{-2} (5.6–18.6 amp ft^{-2}).

MEMBRANES

The membranes occupy a special place in the cost calculation because their lifetime is in general shorter than that of the plant. Their depreciation charges fit none of the three terms of the cost equation accurately. If the lifetime of the membranes is independent of the current

---

*This conclusion is valid not only for electrodialysis but also for any process in which the relevant flux (in this case the ion flux, expressed by the electrical current density) is proportional to the relevant force (in this case the applied voltage). For instance, in reverse osmosis (hyperfiltration, see Chapter 9) the product flow is roughly proportional to the applied pressure. For the general aspects of this "square root law," (Kelvin's law) the reader is referred to the literature on applications of nonequilibrium thermodynamics (see references to Chapter 3 and the Grant and Ireson reference in the literature section of this chapter).

density used, the replacement costs can be included in constant b like other equipment costs, after suitable modification for the shorter lifetime. The lifetime of membranes, however, decreases frequently with increasing current density. The replacement costs per unit water produced are then about the same for one stack operating at maximum current density or two stacks at half that current density. In this case, membrane replacement costs are included in constant c.

Membrane replacement is a major item in the cost of electrodialyzed water. It is very difficult, however, to set accurate estimates of the lifetime of a given membrane for a given application from laboratory experiments alone.

A large number of ion-exchange membranes is now available. Many membranes are reinforced by strong fabrics often made from glass or Dynel* fibers to give them the mechanical strength necessary to withstand pressure differences between adjacent electrodialysis compartments.

A number of relevant membrane characteristics are listed in Appendix 5A. The choice of the most economical membranes for a given application is often made by the plant designer, but it should be kept in mind that many plant designers also manufacture membranes and therefore tend to utilize their own membranes, whose performance they know best.

In principle, membranes have to be only a few molecular layers thick in order to act as electrical ion filters, and there are indications that very thin cell membranes in living organisms do indeed fulfill this function. The necessity for mechanical stability, however, obviates the use of very thin membranes in electrodialysis plants. With membranes 0.1–1.0 mm thick, adequate mechanical properties can be achieved.

The price of membranes varies in accordance with their properties and durability. Most membranes used today have been produced in only limited quantities and hence their prices are relatively high. Present prices of very selective, durable membranes for small plants (10–100 m^3/day) are about \$40–50/m^2. For true mass production, price reduction can be expected.

---

*A copolymer of vinyl chloride and acrylonitrile, Dynel is produced by Union Carbide and Carbon Corp., New York.

ELECTRODIALYSIS

SPACERS

To keep the membranes at a fixed distance and to force the liquids into a tortuous path to enhance mixing, spacers are inserted between the membranes. The liquid compartments in practical electrodialysis units are usually very narrow, the distance between adjacent cation- and anion-exchange membranes being only about 1 mm. This thickness represents a compromise between the electric-power dissipation ("Joule heat") and the pumping power required, which increases and decreases, respectively, with increasing thickness of the compartments. Since the compartments are so narrow, it is obviously impossible to feed and empty each with separate pipes leading to main conduits.

To overcome this problem, the spacers are designed to serve as both conduits and liquid distributors. This principle is illustrated in Fig. 6.2, which shows spacers for the fresh-water and brine compartments in a plane parallel to the membranes. The spacers are made of plastic sheets of thickness equal to the desired cell thickness. Four holes, A, B, C, and D, are punched in each spacer, which match equal holes punched into each membrane and the electrodes. A zig-zag path for the liquids is also punched into each spacer. The two kinds of spacers are identical except for the connections of the liquid path to holes A and C or B and D, respectively. When the alternating layers of membranes and spacers are placed on top of each other and compressed tightly, these holes form liquid conduits.

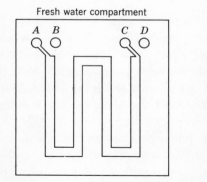 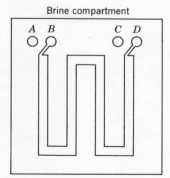

Fig. 6.2. Spacers for electrodialysis units. Principle of liquid-distribution system.

Fig. 6.3. Electrodialysis unit. Membrane stacks and auxiliary equipment are shown. Plants of this type produce 15,000 m³ (4 million gallons) of desalted water from brackish feed water at the island of Corfu, Greece. Electric polarity is periodically reversed for continuous scale control. Courtesy Ionics, Inc., Watertown, Massachusetts.

Feed for the fresh-water compartments, which is either raw water or recirculated partially desalted water, enters these compartments through conduit A but cannot enter the brine compartments. Similarly, feed for the brine compartments enters only the brine compartments through conduit B. Desalted water and brine leave the compartments through separate conduits C and D, respectively.

All practical units use this or similar distribution systems. In many units, however, the central part of the compartment contains a plastic lattice which serves essentially the same purpose as the zig-zag path shown in Fig. 6.2.

An assembled membrane stack is shown in Fig. 6.3.

ELECTRODES

Various materials for electrodes are in use. The positive electrode deteriorates much faster than the negative, because of the particularly corrosive action of oxygen and chlorine evolved at the former. Hence there is a tendency to use platinum-coated electrodes, which are quite resistant (e.g., coated titanium), in spite of the high cost. These platinum coats are very thin, and since each electrode pair serves hundreds of membranes the investment is not prohibitive. Although this protection is particularly important only for the positive electrode, the negative electrode is also made of this material because in many units the current direction is periodically reversed to counteract the effects of polarization and to remove scale deposits.

POLARIZATION AND SCALING

There exists an upper limit of the current density which can be used in electrodialysis units. This limit depends mainly on the composition of the water treated, the flow velocity, and the spacer design and is due to polarization phenomena. In electrodialysis technology, polarization phenomena are defined as effects due to the enrichment and depletion of salt concentration near the membrane surfaces as a result of the passage of the current. The effects of the electrode reactions are also frequently included.

As indicated in Fig. 6.4, enrichment of ion concentration occurs near the surfaces of the membranes facing the brine compartments, whereas there is depletion at the corresponding surfaces facing the fresh water. The salt concentrations near the membranes are therefore different from those in the center of the brine and fresh-water compartments, respectively. This has a threefold effect:

1. Increase of the resistance of the stack and resulting power loss.
2. Formation of scale on the membranes.
3. At high current density, marked changes of the pH in brine and fresh water.

The increase of the resistance is due to the fact that the concentration increase and resulting decrease of the resistance on the brine face of

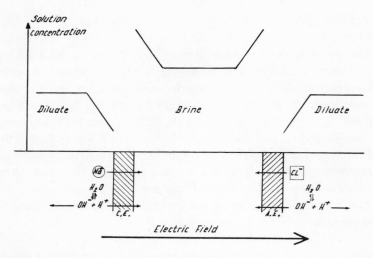

Fig. 6.4. Polarization phenomena in electrodialysis. Lower half of figure schematically shows ion transfer through anion-exchange (A.E.) and cation-exchange (C.E.) membranes and acid-base generation. Upper half shows salt concentration profile in steady state. From Spiegler (1963).

the membranes do not compensate fully for the increased resistances on the other face. This effect is particularly pronounced at high current densities and/or low flow rates, when the rate of turbulent mixing is not sufficient to keep these salt-enriched and depleted layers near the membranes narrow. In fact, if the current density in a stack is maintained constant and the flow rate gradually decreased, a point is reached where the resistance of the stack increases abruptly. This velocity is called the *critical velocity* for the particular current density used.

The increased salt concentrations near the membrane–brine surfaces may induce scale precipitation if the solubility limits are exceeded; calcium carbonate and magnesium hydroxide scales on the anion-exchange membrane–brine interface are quite common. In a plant in Welkom, South Africa, that treated water containing traces of barium and strontium salts, a scale of the sulfates of these metals was found to deposit on the cation-exchange membrane–brine interface. In the electrodialysis of sea water, calcium sulfate scale is common.

Finally, the pH changes are explained by the transport of hydrogen and hydroxyl ions through cation- and anion-exchange membranes, respectively. When the salt concentrations at the fresh-water–membrane surfaces are very low and the current is automatically maintained constant, hydrogen and hydroxyl ions have to participate in the current transport. This *acid–base generation,*[*] as shown in Fig. 6.4, creates an alkaline environment at the anion-exchange membrane–brine interface which favors precipitation of calcium carbonate and magnesium hydroxide.

The formation of a hard, adherent coat of scale on the membranes causes increased electrical and flow resistance and hence power loss and frequently mechanical damage to the membranes. While scale formation is particularly pronounced at high current densities, it does also occur at low current densities although the amount of scale formed per unit amount of product is less. To prevent calcium carbonate scale, hydrochloric or sulfuric acid is often added to the brine steam and also to the rinse liquid for the negative electrode compartment where hydroxyl ions are formed by the electrode reactions. The amounts of acid needed depend on the alkalinity of the raw water. For instance, in a plant in Coalinga, California, where the raw water alkalinity was 140 ppm (as $CaCO_3$) the dosage was 240 g sulfuric acid per cubic meter of fresh water produced (2 lb/1000 U.S. gal). The addition of small amounts of sodium hexametaphosphate (in concentrations as low as a few parts per million) often proves useful to prevent the deposition of scale in the unit. This adds to the cost of the treated water in accordance with the local cost of acid and of sodium hexametaphosphate.

The addition of acid necessitates the use of corrosion-proof equipment and is not desirable because of the hazards involved; hence attempts have been made to operate units free of scale without the addition of acid. Periodic reversal of electrode polarity often leads to dissolution of the scale. In this case, the brine compartments become fresh-water compartments and *vice versa*. Hence the flow connections have to be automatically changed accordingly when the current is reversed.

Waters containing colloidal or coarser solid matter must be pretreated by filtration and/or flocculation. If this is not done, deposits soon

---

[*]*Acid–base generation* was called "water splitting" in the early literature. To avoid confusion with the process of water decomposition into hydrogen and oxygen *gas* (often used in the current energy literature), the term acid-base generation is used here.

form on the membrane, and this leads to problems similar to those encountered with other scales.

ELECTRICAL POWER SOURCE

The direct current needed for electrodialysis is usually produced on site by an a.c. to d.c. converter which has a conversion efficiency of about 90%. In units without diluate circulation, working at constant production rate it is important to keep the current constant because the rate of salt shifting must remain constant. Hence the voltage increases gradually if and when scale deposits. Of course, such constant-current-control devices add to the price of the unit.

In recirculation systems, it is often not necessary to keep the current constant. These systems can work at constant voltage. As desalting proceeds, the resistance rises and both current and salt-shifting rate decrease until a new batch of raw water is processed.

COMPARATIVE EVALUATION OF UNITS

Although the first electrodialytic desalting units appeared on the market only in the early 1950s, a large variety of units of different design are now available. Small compact units which provide 50–100 liters of fresh water per day have been developed. They are stated to consume less than 60 watts and can be hung over a kitchen sink and connected to the regular domestic supply like any other household appliance. Other relatively small and simple units have also been developed for remote farms or households which have a supply of saline water and where some electrical power is available.

Electrodialysis of *sea water* is being carried out in several units but has not been considered economical for fresh-water production in view of the highly concentrated raw water except in special situations. For instance, a plant for electrodialysis of Black Sea water (17,700 ppm dissolved solids) and a Japanese plant, both operated on ships, were described. Development work on reducing the stack resistance by increasing the water temperature (up to 70°C) and decreasing the thickness of the water compartments gives rise to hopes for cost reduction, however.

SELECTED LITERATURE

General Reference on Applied Electrochemistry

Bockris, J. O. M., and Reddy, A. K. N., *Modern Electrochemistry*, 2 vols., Plenum/Rosetta, New York, 1973.

Reviews

Solt, G. S., "Electrodialysis," *Membrane Separation Processes*, P. Meares, ed., Elsevier, Amsterdam, 1976.
Lacey, R. E., and Loeb, S., eds., *Industrial Processing with Membranes*, Chapters 1–3, Wiley, New York, 1972.
Wilson, J. R., ed., *Demineralization by Electrodialysis*, Butterworths, London, 1960.
Spiegler, K. S., "Electrodialysis," *Chemical Engineer's Handbook*, 5th ed., R. H. Perry and C. H. Chilton, eds., McGraw-Hill, New York, 1973, Chapter 17, p. 52.
Passino, R., ed., *Biological and Artificial Membranes and Desalination of Water*, Elsevier, Amsterdam, 1976.

Papers on Specific Aspects Discussed in This Chapter

Belfort, G., and Guter, G. A., "An Electrical Analogue for Electrodialysis," *Desalination* **5**:267 (1968).
Belfort, G., and Guter, G. A., "An Experimental Study of Electrodialysis Hydrodynamics," *Desalination* **10**:221 (1972).
Forgacs, Ch., Ishibashi, N., Leibovitz, J., Sinkovic, J., and Spiegler, K. S., "Polarization at Ion-Exchange Membranes in Electrodialysis," *Desalination* **10**:181 (1972).
Forgacs, Ch., Koslowsky, L., and Rabinowitz, J., "The Desalination of Sea Water by High-Temperature Electrodialysis," *Desalination* **5**:349 (1968).
McRae, W. A., Glass, W., Leitz, F. B., Clarke, J. T., and Alexander, S. S., "Recent Developments in Electrodialysis at Elevated Temperatures," *Desalination* **4**:236 (1968).

Orjerovsky, M., "First Industrial Installation of Electrochemical Desalting of Sea Water," *Proceedings of the Symposium on Salt Water Conversion,* Union Institute for Water Supply, Canalization, Hydrotechnics and Engineering Hydrogeology, Academy of Building and Architecture, Moscow, 1959. In Russian, describes electrodialysis of Black Sea water.

Spiegler, K. S., "Saline-Water Conversion Research in Israel," *Advances in Chemistry Series,* No. 38, American Chemical Society, Washington, D.C., 1963, p. 179.

Tsunoda, Y., and Kato, M., "Compact Apparatus for Sea-Water Desalination by Electrolysis Using Ion-Exchange Membranes," *Desalination* **3:**66 (1967).

Grant, E. L., and Ireson, W. G., *Principles of Engineering Economy,* 4th ed., Ronald Press, New York, 1964. A discussion of economic optimization, by weighing investment costs against electric power costs *(Kelvin's law)* starts on p. 224.

7

FREEZING PROCESSES

PRINCIPLE

Separation of fresh water from salt solutions by freezing is based on the fact that ice crystals which form when salt water is cooled, are essentially salt free. In this respect, ice formation is analogous to distillation, which leads to salt-free vapor although the liquid may hold a high concentration of salts. From a practical standpoint, however, the two processes are different. Distillation is carried out well above ambient temperature and hence the equipment must be designed for minimal heat losses. Only liquid and vapor have to be moved and purified. In freezing methods, the system must be protected against heat gains, or "cold losses," and, in addition to fluids, ice is transported and purified, which is somewhat more complex than the corresponding operations for fluids. On the other hand, the low operating temperature of freezing processes greatly reduces scale and corrosion problems.

Process and plant design are based on the well-known principles and equipment of refrigeration technology, adapted to salt-water purification. It is quite possible that these processes will eventually be competitive with distillation.

Both direct and indirect freezing processes have been developed.* In the former process, water acts as its own refrigerant; i.e., the evaporation of water vapor produces the cold necessary for the formation of ice. In the latter process, a more volatile liquid—e.g., butane—is used as a refrigerant. It has proved possible to maintain direct contact between water and butane, which are almost mutually insoluble, so that it is not necessary to build two different circulation systems *separated by solid surfaces* for heat exchange. A special variant of this method is the hydrate process in which a solid compound of refrigerant and water, rather than ice, is crystallized from sea water.

Direct Freezing

The principle of the direct-freezing process, also known as the "Zarchin process" (or vacuum-flash process), further developed jointly with Colt Industries, Beloit, Wisconsin, is shown in Fig. 7.1. Sea water precooled by heat exchange enters a freezing tower (crystallizer) in which the pressure is maintained between 3 and 4 mm mercury (about 0.005 atm). Rapid evaporation takes place. Since heat is necessary to maintain the evaporation process and since the crystallizer is thermally insulated, the sea water cools and eventually freezes. (The freezing temperature is about $-1.9°C$ for pure sea water and $-3.8°C$ for doubly concentrated sea water, which is the liquid phase in the crystallizer.) In other words, the incoming water splits into vapor and ice, and the heat needed for the evaporation process is supplied by conversion of water to ice. The rate of crystallization achievable in modern freezers is 1–1.5 tons of ice per hour and cubic meter of crystallizer space.

In this conversion, 80 kcal/kg ice has to be withdrawn (143 Btu/lb), whereas evaporation requires about 600 kcal/kg vapor produced. Thus each kilogram vapor can produce 7.5 kg of ice. The ice must be washed because it crystallizes as a fine slush still holding up to 50 weight per cent brine in the space between the crystals. Countercurrent washing proce-

---

*In some of the literature on this process, the term *direct freezing* is used for the production of ice by immersion of refrigerator coils into salt water; while this procedure was studied many years ago, it is not now considered a practical desalination method on the industrial scale.

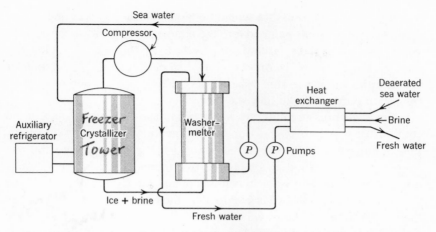

Fig. 7 1. Block diagram of direct-freezing process with vapor compression. Cold sea water evaporates in crystallizer under very low pressure. Ice forms and is brought to washer-melter, where it is contacted with compressed vapor in counterflow, thus forming fresh water. Auxiliary refrigerator adjusts heat balance. Courtesy Colt Industries, Inc., West Hartford, Connecticut, and Beloit, Michigan.

dures have been developed in which the ice rises against a stream of wash water flowing downward. This method is so efficient that only a few per cent of the fresh water has to be wasted for this purpose. Efficient and compact wash columns are now available in which washing is carried out in a pressurized, totally flooded column. The concentrated brine and the fresh water are discharged through a heat exchanger in which the incoming sea water is precooled.

In order to maintain the vacuum in the crystallizer, the water vapor formed by evaporation must be continuously pumped out. If this were not done, the pressure in the crystallizer would soon equal the water vapor pressure of the brine and evaporation would stop; this would also stop the production of ice. Therefore, a large compressor is incorporated in the plant which acts as a pump to remove the vapor from the crystallizer.

In principle, it would be possible to bring the vapor simply up to atmospheric pressure and discharge it and then let the ice melt by itself; however, the amount of power to compress the dilute vapor from 0.005

atm to atmospheric pressure would be large. To save power and equipment, the vapor is not pumped into the atmosphere but into a vessel into which the washed ice is continuously fed. In this vessel, called the *melter,* the pressure is only slightly higher than in the crystallizer because the vapor pressure of ice at its melting point, when in equilibrium with pure water, is only 4.6 mm.

Since the pressure difference between intake and output sides of the compressor is only between 1 and 2 mm mercury, the power consumption is much lower than if the vapor had to be pumped into the air. Moreover, this method accomplishes two things at once. The vapor is not lost but turns into water when contacting the ice in the melter, and the heat for melting the ice is provided by the condensing vapors. Just as the evaporation of 1 kg of water in the freezer causes the formation of about 7 1/2 kg of ice, the condensation of 1 kg of vapor in the melter is sufficient to melt 7 1/2 kg of ice. In fact, too much heat gets into the melter because of the adiabatic action and imperfection of the compressor, which raises the temperature of the vapor considerably, and because of unavoidable heat leakage into the unit. In order to maintain the temperature in the melter at about 0°C and minimize the pressure against which the compressor works, an auxiliary refrigerator unit is necessary.

The melted ice leaves the unit through a heat exchanger in which it precools the incoming sea water.

The performance of the unit depends to a large extent on the compressor. While the latter has to operate only over a small pressure difference, the volume of vapor transported per unit volume of fresh water is very large. At 100°C and atmospheric pressure, the volume of 1 metric ton of saturated water vapor is 1671 m^3. This is the volume to be handled by a compression still operating under these or similar conditions (Fig. 5.11). On the other hand, the corresponding volume for saturated water vapor at 0°C is 206,000 m^3 per ton water. It is seen that the compressor must be designed for moving very large volumes of water vapor over small pressure differences. Rotary fans and blowers are particularly well suited for this type of application. Fans with flexible blades have been developed for this purpose. The development of compressors of this type has benefited not just freezing technology but any kind of low-pressure compression distillation.

FREEZING PROCESSES

To maintain a very low pressure in the freezer it is necessary to deaerate the feed water and to provide a purge system which removes from the vapor space of the freezer those traces of air which have not been removed by the deaerator.

Instead of using a compression-refrigeration system as described, it is also possible to use an *absorption system*. In this process, the vapors are absorbed by a hygroscopic solid or solution which is kept cold during the absorption process by heat exchange with melting ice. The absorbed water vapor is driven off the absorbent by heating in a separate vessel and the absorbent is ready again for the cycle. Absorption-refrigeration systems are quite common in household refrigerators, for instance, the "Servel-Electrolux" type.

Figure 7.2 shows the principle of absorption refrigeration as applied to salt-water purification. Precooled sea water enters the insulated

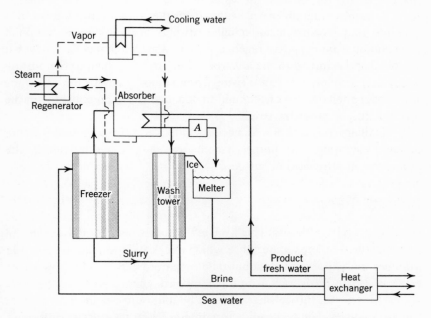

Fig. 7.2. Block diagram of direct-freezing process by absorption. Fully drawn lines show main cycle. Dotted lines show regeneration cycle of absorbent. Auxiliary refrigeration is provided at point A.

freezer where it partially evaporates and where ice is produced as in the process just described. The ice-brine slurry is pumped to a tower where the ice rises slowly while being washed countercurrently with a small percentage of the fresh water produced. The vapor from the freezer is absorbed in a solution containing 50–53% by weight of lithium chloride which has a very low water-vapor pressure. During the absorption process, which proceeds at 12–15°C, heat is given off. It is necessary to maintain the lithium chloride solution at this low temperature because otherwise the water vapor pressure of this solution would rise and it would no longer absorb water vapor from the freezer. In principle, the ice leaving the wash tower could be used as a coolant for this purpose, but it is more convenient to maintain a large amount of fresh water in circulation, which is used to melt the ice and then to cool the solution in the absorber. The ice in the melter is continuously fed into this circulating stream of fresh water. Product is also continuously bled off this stream so that the quantity of water recirculating remains constant.

To remove the absorbed water from the lithium chloride solution and thus keep its concentration high and vapor pressure low, a portion of the solution is pumped to a regenerator containing steam coils in which it is boiled and returned to the absorber after giving up heat to the portion going to the boiler. The vapor rising from the boiler passes into a surface condenser cooled by sea water and the condensate is combined with the recirculating fresh-water stream.

Auxiliary refrigeration is necessary to keep the temperature constant. Deaeration and purge systems are also essential, as in the compression-refrigeration system.

Indirect Freezing

To overcome the drawback of low vapor pressure and density of water at the freezing temperature which necessitate the moving of large volumes of vapor and vacuum-tight apparatus, the indirect-freezing process uses a refrigerant which has much higher vapor pressure than water. Of course, this refrigerant must be immiscible with water so that the two liquids can be readily separated. Different refrigerants, e.g., butane and fluorinated organics such as "Freon 114,"* are used.

*Product of E. I. Du Pont de Nemours & Co., Wilmington, Delaware.

A block diagram of an indirect-freezing process using butane is shown in Fig. 7.3. The boiling temperature of butane at atmospheric pressure is $-0.5°C$, i.e., very close to the freezing point of water. Liquid butane and sea water are introduced into the freezer, which is kept under slightly subatmospheric pressure under which butane boils, deriving its heat of vaporization from the formation of ice. For each ton of butane evaporated, only 1.15 tons of ice is formed because the heat of vaporization of butane at $-3°C$ is merely 91 kcal/kg. The slurry of ice and brine is washed countercurrently with a small amount of fresh water, while most of the butane vapor passes into compressor I, where its pressure is increased to slightly over atmospheric. It is then brought into contact with the ice in the melter, where the reaction is opposite to that in the freezer, i.e., butane vapor turns into liquid and ice melts. Water and butane are now separated in the decanter by virtue of their different densities, 1.00 and 0.60, respectively, and liquid butane is returned to the freezer, while the fresh water leaves the plant after cooling the incoming sea water by heat transfer. Instead of direct-contact melting, the "Freon

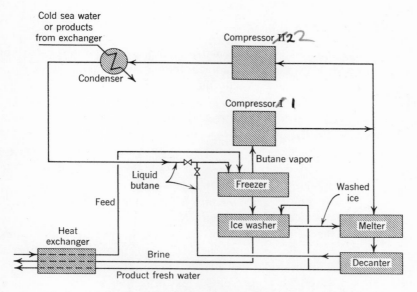

Fig. 7.3. Block diagram of indirect-freezing process. From Karnofsky (1961).

114'' process uses indirect melting. This reduces contamination of the melted ice with the refrigerant.

A small portion of the butane vapor from the freezer is passed to compressor II, which brings it to a higher pressure than compressor I. It then passes to a water-cooled condenser where it gives up heat and liquifies and is then returned to the freezer. This secondary butane cycle with its heat rejection in the condenser represents the auxiliary cooling required to remove heat leakage into the plant and thus maintain the necessary cold temperatures continuously.

It is interesting to compare the power requirements of freezing processes with those of other desalting methods and also to weigh the relative advantages of the direct and indirect freezing processes described in this chapter.

POWER REQUIREMENTS

The minimum power necessary for splitting sea water into fresh water and brine at a given rate of production is roughly proportional to the absolute temperature, as shown in Appendix 1A. This power refers to a completely reversible process in which absolutely no power is unnecessarily degraded to heat. Comparing different *reversible* processes, there should be no difference in their power requirements. All of them start with sea water at the ambient temperature and discharge fresh water and brine at almost exactly the same temperature. It is true that the decisive steps in the various processes are carried out at different temperatures. For instance, the compressors in the direct-freezing and compression-distillation processes described in Chapters 7 and 5 operate at about $-3°C$ and $+101°C$, respectively. If all parts of the equipment operated ideally, however, it would theoretically be possible to recover dissipated power within the plant itself and hence all desalting processes would require the same power input per unit production rate.

In actual practice, the major portion of the power is not utilized for covering the theoretical minimum, but for compensating the various losses in the plant. For instance, in distillation, fresh water and brine leave the plant at considerably higher temperatures than that of the incoming sea water. This represents a loss of heating steam. In freezing processes, product and brine are colder than feed. This requires additional refrigeration capacity and thus entails a heat and/or power loss. When comparing these two types of processes, it is therefore pertinent to ask whether a "lost calorie of heat" is more valuable than a "lost calorie of cold" and then to consider the two processes in the light of the amounts of heat and cold "lost".

To reduce these "losses" to a common denominator, it is logical to consider the performance of absorption refrigerators because they can be driven by steam heat and thus represent a direct conversion of heat into cooling power. If such a refrigerator is powered by saturated steam at 101°C and cooled by water of the ambient temperature 18°C, and if it is to maintain a freezer temperature of $-3°C$, each steam calorie could theoretically remove 2.85 cal from the freezer to the cooling water (see Appendix 6A). This figure refers to an ideally reversible cooling engine which has never been realized in practice. In actual working units, the cooling power is at best 0.7 cal pumped per steam calorie. It follows that the removal of 1 cal from a freezing system at $-3°C$ to compensate for "lost cold" requires more "available" energy than the replacement of 1 cal lost from a still operating at 101°C. If, instead of using an absorption refrigerator, we expand the low-pressure steam of 101°C in a turbine and use the power to drive a compression refrigerator, no better cooling efficiency is obtained. We might say that under most conditions the "loss" of a calorie of cold from a freezing unit is more serious than the "loss" of a calorie of heat from a still, but, on the other hand, fewer calories are likely to leak into a freezing unit than out of a still, because the operating temperature of the former is closer to the ambient temperature. The total heat transfer depends on the design of the unit, thermal insulation, and efficiency of the various components. The actual power consumptions measured reflect all the inefficiencies.

For the actual power consumption of freezing processes, we have still to rely on experiments in relatively small and medium plants and extrapolation of the results to larger units.

WASHING OF ICE

In most freezing processes developed so far, ice is obtained in the form of small crystals. The size of the crystals depends on the conditions of their formation. Normally their characteristic dimensions are about

0.05 mm, but it is now possible to obtain crystals as large as 1.5 mm in a continous process. The ice crystals form a slush with the remaining brine, somewhat similar to wet snow. About half of the weight of the slush is brine and half ice. To move it easily through pumps and lines, some brine is recirculated in most plants.

In general, the ease of washing of slurries increases with the crystal size, but even the small crystals obtained in today's plants can be quite satisfactorily washed by counterflow against fresh water, at the loss of only a few per cent of the latter.

In the wash column (Fig. 7.4), pressures are adjusted such that the ice–brine slush entering at the bottom rises to a location about midway in the column, where the brine diverges outward through peripheral brine–discharge screens, while the solid ice continues to rise. In the upper half of the column, the rising ice is washed by a small amount of counterflowing fresh water which is introduced at the top and also leaves the column through the brine-discharge screens. After being harvested, the ice is melted by condensing vapor (Fig. 7.1), thus producing desalinated water.

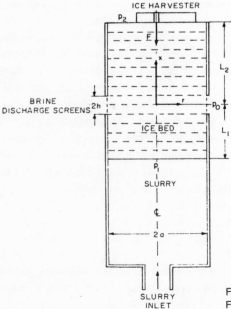

Fig. 7.4. Ice-slurry counterflow washer. From Schwartz and Probstein (1969).

Instead of using a column for the wash process, the Rotocell, a more elaborate device which has proved successful in other washing and extraction operations, has been suggested. This device consists of a large number of open-topped cells which hold the solid and convey it from a feed point around the circle to a discharge point, while it is being washed countercurrently. The separation of ice and brine can also be effected by centrifugation. Wash columns have many practical advantages, however.

HYDRATE PROCESS

When carrying out freezing experiments with different hydrocarbon refrigerants, it is frequently observed that ice is not the only solid phase which forms. For instance, propane is capable of forming a solid hydrate, with 17 molecules of water, although liquid propane dissolves only very small amounts of water. If pure water is gradually cooled in a vessel in which the propane vapor pressure is maintained at 4.9 atm, this hydrate will start to crystallize at +5.0°C. For 3.5% and 7% solutions of sodium chloride, which are similar to normal and doubly concentrated sea water, the crystallization temperatures of propane hydrate are 3.5 °C and 1.6°C, respectively. At the latter temperature, liquid propane (saturated with water) is also stable, since its vapor pressure at that temperature is 4.9 atm. Compare these temperatures to the corresponding freezing temperatures of ice under atmospheric pressure which are 0C°, −1.9°C, and −3.8°C, respectively. It is seen that hydrate formation takes place 5–6°C higher than ice. Hence a process based on the crystallization of propane hydrate could work at a somewhat higher temperature than the other freezing processes.

Some laboratory experiments and preliminary engineering studies for such a process have been made. Its operation is similar to the indirect freezing process. Instead of an ice–brine slurry, a hydrate–brine slurry is moved from the freezer-crystallizer to the wash column and then to a melting vessel. There it contacts propane vapor, which liquifies and decomposes the hydrate into liquid water and propane. The heat of decomposition is about 84 kcal per kilogram of water, almost equal to the heat of melting of ice. The liquid propane is recycled and the pure water withdrawn.

Since the hydrate process would work at a somewhat higher temperature than the regular freezing processes, cold losses to the warmer atmosphere would presumably be less. On the other hand, the equipment would have to withstand higher pressures than in the indirect freezing process with butane as a refrigerant.

SELECTED LITERATURE

The following three Research and Development Reports of the Office of Saline Water, U.S. Department of the Interior, are for sale by the Superintendent of Documents, U.S. Government Printing Office, Washington, D.C. 20402:

Research and Development Report No. 282 (Catalogue No. I1.88:282): Latini, R. G., Rouher, O. S., Agrawal, P. D., and Barduhn, A. J., *The Freezing Process for Desalting Saline Waters*, 1967.

Research and Development Report No. 373 (Catalogue No. I1.88:373): Williams, V. C., Roy, C. L., Smith, H., Jr., and Battle, O. B., *Development of Propane-Hydrate Desalting Process*, 1968.

Research and Development Report No. 451 (Catalogue No. I1.88:451): Consie, R., Darling, R., Emmermann, D., Fraser, J., Johnson, W., Koretchko, J., and Torvbraten, F., *Vacuum-Freezing Vapor-Compression Process: One and Five Million Gallon per Day Desalting Plants*, 1969.

Zarchin, A., *Method for Separating the Solvent from the Solute of Aqueous Solutions*, U.S. Patent 2,821,304 (Jan. 28, 1958).

Zarchin, A., *Process and Apparatus for Sweetening Sea Water*, British Patent 806,727 (Dec. 31, 1958).

Karnofsky, G., "Saline Water Conversion by Freezing with Hydrocarbons," *Chem. Eng. Progr.* **57**:42 (January 1961).

Bosworth, C. M., Barduhn, A. J., and Sandell, D. J., "A 15,000 Gallon per Day Freeze-Separation Pilot Plant for Conversion of Saline Waters," *Advances in Chemistry Series*, No. 27, American Chemical Society, Washington, D.C., 1960, p. 90.

Shwartz, J., and Probstein, R. F., "Experimental Study of Slurry Separators for Use in Desalination," *Desalination* **6**:239 (1969).

Knox, W. G., Hess, M., Jones, G. E., and Smith, H. B., "The Hydrate Process," *Chem. Eng. Progr.* **57**:66 (February 1961).

8

ION EXCHANGE

PRINCIPLE

Complete removal of salts from water by ion exchange has been practiced since the middle 1930s and is a routine water treatment operation for high-pressure boilers which require completely salt-free water. This process is called "demineralization," since it merely removes the electrolytes, which are largely of mineral origin. In its conventional form, the process consumes chemicals in amounts roughly proportional to the amounts of salt removed. It is competitive with other desalting methods only for water containing relatively low concentrations of salts. Ion-exchange methods are important where water of extremely low salt content and electrical conductance is needed, e.g., in the manufacture of television tubes. These applications are not within the primary scope of this book; however, for the sake of completeness and because ion exchange is used occasionally to convert salt water into drinking water, a brief description is given here. For more detailed information, the reader is referred to monographs dealing with *ion exchange* exclusively.

The principle of ion-exchange desalting is shown in Fig. 8.1. Raw water is passed through a column containing the active form of a solid cation exchanger, which is an organic resin containing hydrogen ions, H^+, and capable of exchanging them against the positive ions contained in the raw water. If the resin and the water were merely mixed in a tank instead of the water being passed through the column, the exchange would only be partial. In the column, however, the water first exchanges ions with the top layer of the resin, which it exhausts partially of

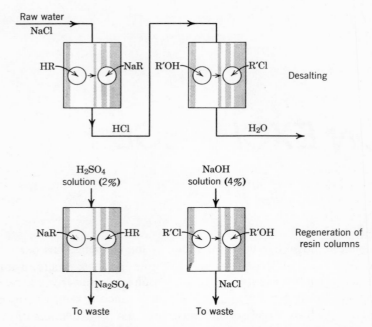

Fig. 8.1. Principle of ion-exchange desalting. HR and NaR are the active and exhausted forms, respectively, of the cation exchanger. R′OH and R′Cl are the corresponding forms of the anion exchanger.

hydrogen ions and then moves slowly down the column where it contacts fresh resin and exchanges those positive ions which were not exchanged in the top layer. Thus when the water leaves the first column, all the original positive ions have been replaced by hydrogen ions and the water has in fact turned into a dilute solution of an acid. Figure 8.1 shows the ion-exchange reactions for sodium ions. Analogous reactions take place with calcium and magnesium ions.

Most raw waters contain some bicarbonates, which are turned into carbonic acid by this procedure and can be removed by aeration. (This step is not shown in Fig. 8.1.) The acid water then passes into an anion-exchange column where another resin exchanges all negative ions against hydroxyl ions, OH^-. Thus all ions originally present are changed into H^+ and OH^-, respectively, which combine to form pure H_2O.

After a few hours of desalting operation, the resins usually have exchanged most of the H^+ and OH^- ions they contained at the start. It is then necessary to regenerate them to the active form. This is done by passing moderately concentrated solutions of acid and hydroxide through them, respectively. At first sight, it might seem paradoxical that in the desalting step of the process cations and anions from the water are replaced by hydrogen and hydroxyl ions, respectively, from the resins, while in the regeneration step exactly the opposite occurs. One reason for this is the fact that the relative preference of the resins for certain ions decreases with increasing concentration. When in contact with natural water or some other very dilute solution, a cation exchanger prefers calcium and magnesium ions rather than hydrogen ions; yet it releases most of the calcium and magnesium when in contact with a solution containing a high concentration of hydrogen ions. But even if this were not so, it would eventually be possible to remove all calcium and magnesium from the cation-exchange column by continuous passage of a dilute acid solution through it for a sufficiently long time.

After regeneration, the resins are rinsed and are now active again.

Instead of using two separate columns, we can use a single column containing a mixture of cation- and anion-exchange resins. In this case, provision is made for hydraulic separation of the two kinds of resins when they are exhausted, followed by separate regeneration and washing and then remixing by compressed air. In the hydraulic separation the lighter anion-exchange resin forms the top layer.

Ion-exchange *desalting* is different from ion-exchange *softening*, which is the exchange of the hardness-forming calcium and magnesium ions in water for sodium ions. Thus softening is a cation-exchange process whereas complete desalting requires both cation and anion exchange; "hard" waters often require presoftening by ion-exchange or other methods prior to desalting proper. The same cation-exchange resins are used in the two processes and the construction of the columns is similar, but the desalting columns need more effective corrosion-protective coatings, because they are in contact with acid solutions.

While the operation of ion-exchange columns is mechanically not very complex, it is worth noting that the design of efficient column processes is rather complicated, because not only does the composition of resin and solution vary from top to bottom of the column, but also, at any particular depth of the column, the composition of solid and liquid changes with *time*. Compare this with "steady-state" processes, e.g., heat transfer, in which a cold and a hot fluid continuously flow through spaces separated by solid heat-transfer walls. In these units, the temperatures of both fluids vary with position but (in the steady state) remain constant in any particular location of the unit. The design theory of ion-exchange *columns* is much more complicated, especially when the fluid contains numerous positive ions, e.g., sodium, calcium, and magnesium, or numerous negative ions, e.g., chloride, sulfate, and bicarbonate. In fact, the development of ion-exchange technology has stimulated important developments in design methods for columnar separation methods in general, wherever separations of different dissolved materials are attempted by passing the solution through columns of solids which adsorb the different dissolved species with different strength and/or at different rates. For historical reasons, such separations are often called *chromatographic separations*. They are used not just for the softening of hard waters but for many other purposes, e.g., for the separation of rare-earth ions and other dissolved species which have similar properties and are hard to separate by other methods.

Although ion exchange is usually performed intermittently, in stationary columns, as shown schematically in Fig. 8.1, continuous ion-exchange processes have been developed, in which the solid resin is moved countercurrently to the solution to be demineralized. This has the advantage of requiring a smaller resin inventory than when resting columns of resin are used, in which a substantial portion of the resin does not appreciably contribute to the demineralization process part of the time. (For instance, right after regeneration, only the upper portion of the column effectively participates in the ion-exchange process, while the rest of the column does not contribute much). On the other hand, units for *continuous* exchange are mechanically more complicated than resting columns. Contrary to the situation in countercurrent flow units for fluids, e.g., gas–liquid absorption units used in the petroleum industry, the *solid* nature of the resins and the attendant danger of excessive breakage of the moving solid particles complicate the construction of these units.

RESIN CAPACITY

A good ion-exchange resin should have a high exchange capacity, i.e., a large amount of exchangeable ions per unit amount of resin, so that desalting can be carried out for a reasonably long time before the operation has to be interrupted for regeneration. The capacity is often listed as milliequivalent exchangeable ions per gram dry resin and is about 5.0 on this scale for commercial cation-exchange resins (corresponding to 115 g sodium ions absorbable per kilogram dry resin) and 3.0 for commercial anion-exchange resins (corresponding to 106 g chloride ions per kilogram).

In many publications, the capacity is given in kilograins calcium carbonate per cubic foot of column space. In other words, this unit designates the amount of salt (as calcium carbonate) which undergoes ion exchange with the amount of resin contained in 1 ft^3 of a loosely packed column. The amount of salt is expressed in terms of calcium carbonate (although calcium carbonate is insoluble), i.e., each gram equivalent salt is taken as 50 g. On this scale, the resin capacities are roughly 40 for the cation exchanger and 24 for the anion exchanger.

The total exchange capacity of a resin column is never utilized because desalting is not complete when the resin nears exhaustion. In practice, about 70% of the total capacity is used before regeneration.

AMOUNTS OF CHEMICALS FOR REGENERATION

The ratio of the equivalent weights of salt, sulfuric acid, and sodium hydroxide is 58.4:49:40. In other words, for each kilogram of salt removed, 0.84 and 0.69 kg of these two regenerants are necessary. In practice, excess of 50–300% of regenerant is used, depending on the degree of salt removal desired. Very roughly, at least 1.5 kg each of pure acid and base are needed for each kilogram of salt removed. For instance, if the feed contains 350 ppm salt, the daily requirements of a desalting plant of daily capacity 1000 m^3 are roughly 0.5 ton of acid and 0.5 ton of caustic soda. If the salinity is 3500 ppm, more than 5 tons of each chemical is needed daily.*

Because of the large amounts of chemicals necessary for regeneration, conventional ion-exchange demineralization is uneconomical *for the treatment of highly saline water*. Attempts have been made to utilize the regenerants more completely by using special ion-exchange resins and/or regeneration procedures. For instance, the sodium form of "weak-acid" cation-exchange resins, in which the cation-exchanging

*Costing roughly $700 or 70¢ per cubic meter of fresh water (in 1976).

groups are carboxylic can be regenerated at better efficiency and/or with weaker acids than the more common "strong-acid" cation exchangers, in which the cation-exchanging groups are sulfonic. This implies that the "weak-acid" resins have a greater tendency to hold (and a lesser tendency to release) hydrogen ions than the "strong-acid" resins. As a result, the weak-acid *cation-exchange* resins remove sodium ions from saline water in the desalting step to a lesser extent than the strong-acid resins, but unless very pure water is desired this penalty is often considered minor as compared to the saving of acid in the regeneration step. Similarly, weak-base *anion-exchange* resins (containing weakly basic amine-exchange groups) require less regenerant, but are less efficient anion removers, than strong-base anion-exchange resins (which contain strongly basic quaternary anion-exchange groups). The weak-base anion-exchange resins can often be regenerated with relatively inexpensive bases, e.g., ammonia.

Various combinations of "weak" and "strong" exchangers have been successfully worked out for the desalination of brackish waters, with the anion-exchange resin column sometimes preceding the cation-exchange column. Some processes use more than a single pair of cation and anion exchangers in succession. To find if any of these processes is economically preferable to membrane processes, each brackish-water case has to be analyzed on its own merits, keeping the nature of the salts dissolved in the water well in mind. In general, economic prospects for these methods should be analyzed in particular when the salt concentration is less than a few thousand parts per million.

SPECIAL MODIFICATIONS OF ION-EXCHANGE DESALTING

During World War II, the Permutit Companies in the United States and in England developed a sea-water desalting kit for the production of small amounts of drinking water for pilots shot down and stranded in liferafts. The major constituent of this kit is a cation-exchange material containing silver ions, which on contact with sea water are exchanged for the major portion of the cations in sea water. The silver ions do not stay in solution, since they are precipitated by the chloride ions as insoluble silver chloride. The ion-exchange material is compressed into

ION EXCHANGE

Fig. 8.2. Ion-exchange desalting plant. A unit of this type provided part of the drinking-water supply for the town of Eilat on the Red Sea. The raw-water source was brackish ground water. Courtesy Ayalon and Etzioni, Ltd., Haifa, Israel.

briquets, which also contain (1) a swelling clay which acts as a disruptor to promote fast mechanical breakup upon contact with sea water, (2) some silver oxide which reacts with the salts in sea water to form insoluble magnesium hydroxide and more silver chloride, and (3) activated carbon for taste improvement. All these chemicals plus the containers occupy less than one-sixth of the volume of the drinking water they can produce. The briquets and sea water are contacted in a plastic bag in which the ion-exchange and precipitation reactions take place. The treated water still contains several thousand ppm salt but is adequate for emergency use. It is withdrawn through a built-in filter. The kit is shown in Fig. 8.3.

Fig. 8.3. Sea-water desalting kit for emergencies. Courtesy Permutit Co., Inc., New York.

To save chemicals, various modifications of conventional ion-exchange desalting seem possible. Instead of exchanging the salt-water ions against hydrogen and hydroxyl, the salts in solution can be converted into ammonium bicarbonate in a mixed-resin bed. The resulting solution is then heated, causing the evaporation of the bicarbonate with some water. The condensate is a 13% bicarbonate solution which is used for regeneration of the resins. Thus, instead of chemicals, heating steam for the bicarbonate distillation is required. The promising "Sirotherm" process uses ion-exchange resins that can be regenerated by steam or hot water, which contains higher hydrogen and hydroxyl concentrations than cold water. At low temperatures the resins adsorb salt ions, while at high temperatures regeneration takes place. Thus the required available energy ("exergy") for partial salt-water separation is derived from a temperature cycle rather than from the use of acid and base.

SELECTED LITERATURE

General

Helfferich, F., *Ion Exchange,* McGraw-Hill, New York, 1962.
Nachod, F. C., and Schubert, J., *Ion-Exchange Technology,* Academic Press, New York, 1956.

Detailed Theory of Column Operation

Helfferich, F., and Klein, G., *Multicomponent Chromatography: Theory of Interference,* Marcel Dekker, New York, 1970.

Continuous Ion Exchange

Probstein, R. F., "Desalination," *American Scientist* **61(3)**:280 (1973). This article has a photograph of a continuous-moving-bed unit.

Use of Weak-Acid and Weak-Base Resins

Kunin, R., and Vassiliou, B., "New Deionization Techniques Based upon Weak-Electrolyte Ion-Exchange Resins," *Ind. Eng. Chem. (Process Design Dev.)* **3**:404 (1964).

Use of Evaporator Brine for Regeneration of Ion-Exchange Softeners

Sephton, H. H., and Klein, G., "A Method of Using Irrigation Drain Water for Power-Plant Cooling," *Proceedings of the First Desalination Congress of the American Continent* (Mexico City), Vol. 1, Elsevier, Amsterdam, 1976, p. III-2.

Thermal Regeneration of Ion-Exchange Resins

Weiss, D. E., Bolto, B. A., McNeill, R., Macpherson, A. S., Siudak, R., Swinton, E. A., and Willis, D., "The 'Sirotherm' Demineralization Process—An Ion-Exchange Process with Thermal Regeneration," *Proceedings of the First International Symposium on Water Desalination* (Washington, D.C.), Vol. 2, 1965, p. 3, for sale by Superintendent of Documents, Washington, D.C. 20402.

Bolto, B. A., "Sirotherm Desalination—Ion Exchange with a Twist," *Chem. Technol.* **5**:303 (1975).

Miscellaneous

Permutit Co., Inc., "Permutit Emergency Desalting Kit," New York, 1958.

Stewart, P. B., "Sea Water Demineralization by Ammonium Salts Ion Exchange," *Advances in Chemistry Series,* No. 27, American Chemical Society, Washington, D.C., 1960, p. 178.

Spiegler, K. S., Juda, W., and Carron, M., "Counterflow Regeneration of Cation Exchanger in Partial Demineralization of Brackish Waters," *J. Amer. Water Works Assoc.* **44**:80 (1952).

9

REVERSE OSMOSIS (HYPERFILTRATION)

In all filtration processes, liquids or gases are separated from solid or liquid impurities which form a separate phase. Clay particles can thus be separated from water or tar particles from cigarette smoke. For many years it was thought that no filter could retain salt while passing water, because a salt solution represents a single phase. Experiments in the late 1920s and early 1930s, however, demonstrated small filtration effects in very dilute solutions with collodion membranes.

As a result of growing interest in water desalting, the problem was again attacked in the 1950s. By using extremely dense filtering media, almost complete removal of salt even from concentrated solutions was achieved with special synthetic membranes. This process is called *hyperfiltration** or *reverse osmosis*.

Membranes which are permeable to water but not to salt are called *semipermeable*. The principle of the process is illustrated in Fig. 9.1.

When a semipermeable membrane is placed between sea water and pure water which are both at the same pressure, diffusion of fresh water into sea water will occur because of nature's tendency to equalize concentrations. This process, called *osmosis*, is exactly the opposite of the desired action, namely, the transfer of water from the salt solution

---

*In the older literature, the process is called *ultrafiltration*, a term coined by colloid scientists for the filtration of very small particles.

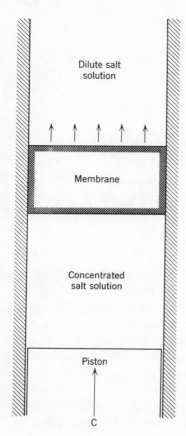

Fig. 9.1. Hyperfiltration through a membrane.

into the fresh-water reservoir. To effect the process in the desired direction, pressure must be exerted on the salt water. The minimum pressure necessary can be calculated from the minimum work for desalting for any reversible process, i.e., 0.70 kwhr for the production of 1 m³ of fresh water from sea water at 25°C, as shown in Chapter 3. Since 1 kwhr equals 3.55×10^4 liter-atm or 35.5 m³ atm, it follows that the work expended by pushing 1 m³ of water through a pressure differential of $0.7 \times 35.5 = 24.8$ atm is equivalent to 0.7 kwhr. In other words, the pressure to be exerted on the sea water to filter it through the membrane and thus convert it to fresh water must be at least slightly larger than 24.8 atm, which is defined as the osmotic pressure of sea water; hence the name *reverse osmosis*.

In practice, pressures must be higher to achieve reasonable filtration rates, and it must be taken into account that the salt concentration in the sea water increases as the process proceeds and hence the osmotic pressure increases too. For doubly concentrated sea water the osmotic pressure is roughly doubled, but the increase of osmotic pressure at high concentrations is not strictly linear. For instance, Dead Sea water has a much higher osmotic pressure than expected from its salt concentration alone.

The crux of this method is the preparation of suitable membranes. This problem has been attacked from different directions. Films of plastic materials such as polyvinyl alcohol or cellulose acetate were prepared and a systematic study of the relation between their composition and salt-filtering ability carried out. Another approach starts from the well-known fact that when an ion-exchange material is in contact with a salt solution, it takes up water in preference to salt. This phenomenon is known as the *Donnan effect* and has led to the use of ion-exchange membranes for hyperfiltration.

Many units contain membranes made from modified cellulose acetate. These membranes combine high salt rejection (95–99%) for the desalination of brackish water and relatively high flow rates, i.e., 6–12 liters per square meter of membrane surface, day, and atmosphere excess pressure, which corresponds to 8.4–16.8 lb water per square foot membrane surface, day, and each 100 psi excess (i.e., over osmotic) pressure. They were developed in Los Angeles in the late 1950s by S. Loeb and S. Sourirajan, who worked at that time for the Saline-Water Conversion Research Program of the University of California. While membranes which can separate salt from water had been known before, the relatively high flow rates—believed to be due to the fact that the separation is caused by an extremely thin dense skin, while the rest of the membrane is quite permeable—made the process practical for industrial use. These membranes are made by casting a solution containing cellulose acetate and other ingredients on a suitable support, submerging it in water, and then curing the resulting membrane at a given temperature. The higher this temperature, the higher the salt rejection and the lower the membrane permeability.

Figure 9.2 illustrates some practical reductions to practice of the principle of reverse osmosis. In several types of units (Figs. 9.2a–c), saline water flows through cylindrical membranes backed by solid or

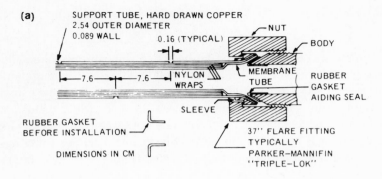

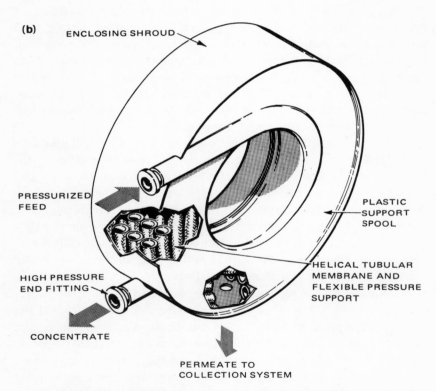

Fig. 9.2. Types of hyperfiltration units. (a, b, c) Tubular units, Loeb design (*Desalination*, **1**:35, 1966), helical tubular segment (Philco-Ford Co., Newport Beach, California, 1972), "sand-log" design (Membrane System Project, Heat Transfer Division, Westinghouse Electric Corp., Pittsburgh, Pennsylvania, 1973), respectively. In the last unit, the membrane tubes are embedded in sand. (d) Spiral-wound unit (United Oil Products-ROGA, San Diego, California, 1966). (e) Hollow-fiber unit, "Permasep" (E. I. Du Pont de Nemours Co., Wilmington, Delaware, 1967). From Harris *et al.* (1976).

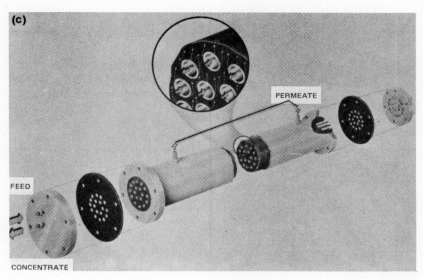

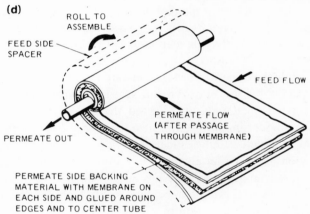

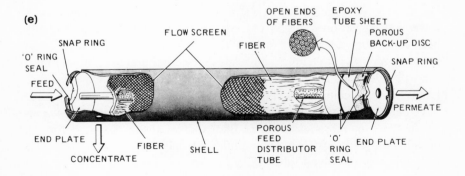

flexible tubes. The desalted water ("permeate") collects in a porous backing between membrane and support and flows to exit ports. Tubular units were used in a plant for the treatment of agricultural waste water in Firebaugh, California.

Tubular units can be easily cleaned, but the high volume/surface ratio is necessarily relatively large. Because of the high pressure (20–100 atm) exerted on the salt water, it is desirable to have, for any given membrane surface, the lowest possible volume to contain the high-pressure fluid. The spiral-wound and hollow-fiber designs were developed with this aim. In *spiral-wound units,* membranes and backing are wound similar to a jellyroll around a central, perforated tube which collects the product. Saline-water feed flows through separate channels in the direction shown in Fig. 9.2d. The membranes are usually also made from modified cellulose acetate, and separate modules can readily be connected in series or in parallel, as indeed can tubular modules. In hollow *fine-fiber units* (Fig. 9.2e), very large numbers of hollow fibers, thinner than human hair having their ends potted in epoxy resin, are held in a pressure vessel. Pressurized saline water circulates on the outside of the fibers, while the hyperfiltrate flows toward the open ends of the fibers held in position by the epoxy resin. Desalted water emerging from millions of open fiber ends is collected there. The hollow fibers are made—by methods similar to those developed in the textile fiber industries—from modified cellulose acetate. They can also be made from special "nylon"-type materials (e.g., in the unit shown in Fig. 9.2e) or other special polyamide-type materials, all of which have been found to separate salt from water and exhibit reasonably high flow rates.

Inasmuch as the separation of salt from water by this method is a process of molecular filtration, one should expect that for active membrane layers of equal thickness, increased permeability is correlated with lower salt rejection. This is indeed usually the case. Several highly permeable films deposited from different materials, e.g., humic acids on porous supports ("dynamic membranes"), were found to exhibit moderate salt rejection, especially for waters of low salinity. On the other hand, modified cellulose acetate membranes and polyamide membranes have been developed for hyperfiltration of *sea water,* but the permeability of these membranes is lower than that of those designed for the desalination of brackish waters containing only a few thousands parts per million of dissolved salts.

Because of the molecular-filter character of hyperfiltration membranes, even relatively small particles can plug the membranes, thus reducing their production rates, and often also their salt rejection. Hence feed waters usually have to be treated before the membranes are exposed to them. Most surface waters are too contaminated for direct hyperfiltration, and even clear ground waters often contain too high a concentration of calcium and/or other scale-forming ions, so that at least partial softening is necessary to prevent the formation of insoluble precipitates at the membrane surfaces. Conventional water-treatment methods (including those discussed in Chapter 4) are used to pretreat the raw water. Clarification by lime-softening and coagulation followed by conventional filtration is often practiced. Chlorination serves to oxidize and thus eliminate organic matter and microorganisms, but *excess* chlorine has often to be removed (e.g., by addition of small amounts of bisulfite to the chlorinated raw water), because many hyperfiltration membranes are sensitive to the oxidizing action of chlorine. [The adequacy of pretreatment can be measured by an empirical filter-plugging test, using filter paper of nominal pore size 0.45 μm (0.00045 mm) from which the "plugging index" is determined.] All these pretreatment methods, most of which were developed a long time ago for the treatment of municipal and industrial waters, add substantially to the total cost of desalination by reverse osmosis.

HYPERFILTRATION OF SEA WATER

Sea water can be desalted by two-stage hyperfiltration, in which a portion of the sea water is first passed through one membrane and the hyperfiltrate is subjected to reverse osmosis again. This is done because frequently the salt concentration of the first hyperfiltrate is too high for practical use of the product. Membranes have been developed, however, to reject a very high percentage of the salt, so as to produce a low-concentration hyperfiltrate in a single "pass" (stage). Such membranes reject more than 99% of the salt. Replacing conventional cellulose acetate, used for the production of brackish-water hyperfilters, by cellulose diacetate-triacetate blends (or pure triacetate) and/or choosing the curing temperature of the membranes judiciously, rejections of more than 99% can be obtained. Certain polyamide membranes which have been prepared also exhibit salt rejections exceeding 99%. These materials have been prepared in the form of hollow fine fibers (Fig. 9.2.e), but polyamides can also be incorporated in flat films which are the hyperfilter sheets in spiral-wound modules (Fig. 9.2d). They can be formed by in-place polymerization of the hyperfiltrating film [e.g., a cross-linked poly(ether/amide)] on a porous support membrane (e.g., a polysulfone). This is a very useful technique, because the active film and the support material can be chosen more or less independently, while in the modified cellulose-acetate membranes, active film and porous support have similar chemical composition, albeit different porosity. Moreover, the thin-film composite membranes can often be stored in the dry state while the original modified cellulose acetate membranes cannot.

Because the osmotic pressure of sea water is much higher than that of brackish water, *sea-water* hyperfiltration modules have to withstand higher pressures. They are often encased in strong fiberglass-reinforced plastic vessels which are not subject to corrosion. In 1976, both hollow-fine-fiber and spiral-wound modules were commercially available. The optimum size of modules has yet to be determined. Single modules available in 1976 could produce up to 20 m^3 of hyperfiltrate per day at brine pressures from 50 to 70 atm. They are cylindrical (outer diameter up to 0.2 m). In principle, it is possible to increase the fresh-water production rate by application of higher pressures, but the increased cost and weight of the units and the flux reduction due to membrane compaction, which is particularly pronounced with flat-sheet membranes, place an upper limit on useful pressures. It should be possible to recover a major portion of the mechanical energy of the high-pressure brine by discharging it to the atmosphere via turbines. Even without these turbines, the practical requirement of mechanical or electrical energy for sea-water hyperfiltration is only about 10 kwhr per ton of product, as compared to the theoretical requirement of 1 kwhr per ton for an *infinitely* slow process (Appendix 1A). It is doubtful that any process producing fresh water at *practical* flow rates which do not necessitate equipment of enormous size could operate at better power efficiency.

When one compares the fuel requirements of a multistage flash-distillation plant with the fuel necessary to produce the electrical power for a hyperfiltration plant of the same production rate, one finds that the fuel requirement for hyperfiltration is of the same order, or even less than for multiple-stage flash distillation. The choice between the two methods depends on many factors, in particular, local cost and availability of fuel and power, but it is of interest that under favorable circumstances sea-water *hyperfiltration* possibly with in situ membrane manufacture offers an economically viable alternative to *distillation*.

In the mid-1970s it seems that both the synthesis of efficient hyperfiltration membranes and the design of molecular-filtration units for their use could still be improved, but this technology has made giant strides in the last decade. Much remains to be learned about operating procedure of hyperfiltration plants. More operating experience about membrane methods in general will no doubt come from a large plant planned for reducing the salinity of a tributary of the lower Colorado River. This plant is to be built near Yuma, Arizona, under the auspices of the Bureau of Reclamation, U.S. Department of the Interior. It is to produce somewhat more than 100 million gallons (about 400,000 tons) of low-salinity water per day.

MISCELLANEOUS

A number of desalting methods have been tested in the laboratory and for various reasons have not developed on a larger scale. Even though these methods are not deemed competitive with the older ones, their study focused attention on the bottlenecks, which might be overcome at a later stage, as has often happened in the past when processes were placed on the shelves for lack of an essential material or process development and were later revived and applied when that material or process became available. Hence it is important to know about these processes, and therefore some of them are briefly discussed in the following.

Solvent Extraction

It has been found that certain organic solvents can extract more or less pure water from saline water in one temperature range and release the water at a different temperature. For instance, triethylamine dissolves about 30% by weight of water at 20°C and only 2.5% at 50°C.

Hence, if saline water and this solvent are contacted and brought to equilibrium at 20°C and the solvent phase then heated to 50°C, most of the water extracted precipitates at 50°C as a separate layer which can be withdrawn. This water still contains about 3% of amine which can be stripped by steam in a column in which water and steam flow countercurrently.

The principle of a solvent extraction unit is shown in Fig. 9.3. In the past, such units have been operated only on the bench scale, but

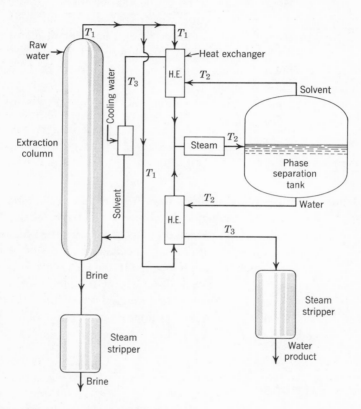

Fig. 9.3. Schematic of solvent extraction process for saline-water conversion. Cold salt-water and solvent flow countercurrent through extraction column. Steam heats wet solvent to effect water precipitation. Steam strippers recover traces of solvent. Components marked "H.E." are heat exchangers. *T* designates temperature. From Hood and Davison (1960).

production plants would probably follow a similar scheme. The heart of the unit is the extraction column in which cold salt water and solvent are contacted in countercurrent flow. The water-laden solvent is heated by heat exchange with fresh water and recycle solvent from the phase separation tank and, finally, by steam. The rise in temperature causes precipitation of water in the phase-separation tank by gravity.

Both fresh water and waste brine still contain a small amount of solvent which is recovered by steam injection in separate columns (steam strippers). Moreover, the solvents used in the past are not completely selective for water. They also extract some salt and hence separation is not complete. Some of this drawback can be overcome by new solvents. Perhaps solid materials can take the place of liquid solvents. For instance, dry ion-exchange resins extract water in preference to salt. The partially desalted water can then be squeezed out or perhaps separated by temperature change. Other solids called *snake cage resins* remove salt in preference to water at room temperature and can be regenerated at high temperatures. It is not particularly important whether extraction takes place at a lower temperature than phase separation or *vice versa*.

Critical-Pressure Distillation

In critical-pressure distillation, sea water is distilled at a very high temperature and pressure, near the critical point of water (374°C, 218 atm) where the densities of pure liquid water and water vapor are identical and equal to 0.4 g/ml. Under these conditions, the heat of evaporation of sea water is extremely small. This is an advantage compared to distillation at or near atmospheric pressure, where the heat of evaporation is very large. Of course, it is necessary to pump the incoming sea water to the very high pressure and heat it to the high temperature at which distillation is to take place. Part of the pumping energy can be recovered by discharging product and blowdown through turbines coupled to the feed pump. Most, but not all, of the necessary heat is supplied to the feed water by countercurrent heat exchange against product and brine. On the other hand, the cost of the high-pressure equipment which must be heavily insulated against heat losses is high and the problems of scale formation and corrosion at these high temperatures have in the past held up the development of this process. It has not been carried into practice, although preliminary work on some of its phases and feasibility studies have been carried out.

Osmionis and Other Methods

Various other processes have been tried and laboratory tested, e.g., zone freezing, electrolysis with electrically induced absorption on porous electrodes, methods for the *chemical* separation of salts, water absorption by inorganic hydrating agents followed by release at higher temperatures, and even biological methods involving the use of algae. Another process which has been suggested is *osmionis,* a name coined from "osmosis" and "ions" and referring to a membrane process in which the diffusion of ions from a very concentrated salt solution into a brackish-water compartment causes desalting of another portion of the brackish water by suitable compartment formation with permselective membranes. Information in these processes can be found in the early annual progress reports of the Office of Saline Water, United States Department of the Interior.

SELECTED LITERATURE

Harris, F. L., Humphreys, G. B., and Spiegler, K. S., "Reverse Osmosis (Hyperfiltration) in Water Desalination," Chapter 4 in *Membrane Separation Processes,* P. Meares, ed., Elsevier, Amsterdam, 1976.

Sourirajan, S., ed., *Reverse Osmosis,* Natl. Res. Council Canada, Ottawa, 1977.

Merten, U., ed., *Desalination by Reverse Osmosis,* M.I.T. Press, Cambridge, Mass., 1976.

Channabasappa, K. C., and Strobel, J. J., "Status of Sea-Water Reverse Osmosis Membrane Process Technology," *Proceedings of the International Symposium on Fresh Water from the Sea* (Alghero), Vol. 4, 1976, p. 267, for sale by A. A. Delyannis and A. E. Delyannis, Tsaldari St. 34, Athens-Amaroussion, Greece. This article also deals with the definition of the "plugging index" and with description of the equipment for its measurement.

Dow Chemical Co., *Permselective Hollow Fibers and Method of Making Them,* U.S. Patent 3,423,491 (1966).

E. I. DuPont de Nemours & Co., *Reverse-Osmosis Separations Using a Treated Polyamide Membrane,* U.S. Patent 3,551,331 (1971).

Taylor, I. G., and Haugseth, L. A., "Yuma Desalting Plant Design," *Proceedings of the First Desalination Congress of the American Continent* (Mexico City), Vol. 1, Elsevier, Amsterdam, 1976, p. V-4.

Lazare, L., *Process for Reducing Salt Content of Salt-Containing Water,* U.S. Patent 3,386,913 (June 4, 1968). Solvent extraction.

Hood, D. W., and Davison, R. R., "The Place of Solvent Extraction in Saline-Water Conversion," *Advances in Chemistry Series,* No. 27, American Chemical Society, Washington, D.C., 1960, p. 40.

McKelvey, J. G., Spiegler, K. S., and Wyllie, M. R. J., "Salt Filtering by Ion-Exchange Grains and Membranes, *J. Phys. Chem.* **61**:174 (1957). Early experiments on use of synthetic ion-exchange membranes for hyperfiltration; absorption of fresh water by ion-exchange granules and recovery by squeezing.

Hatch, M. J., Dillon, J. A., and Smith, H. B., "Preparation and Use of Snake-Cage Polyelectrolytes," *Ind. Eng. Chem.* **49**:1812 (1957).

Riley, R. L., Fox, R. L., Lyons, C. R., Milstead, C. E., Seroy, M. W., and Tagami, M., "Spiral-Wound Poly(Ether/Amide) Thin-Film Composite Membrane Systems," *Proceedings of the First Desalination Congress of the American Continent* (Mexico City), Vol. 1, Elsevier, Amsterdam, 1976, p. II-1.

Glueckstern, P., and Greenberger, M., "Field Tests and Engineering Evaluation of Commercial Reverse-Osmosis Units for Brackish-Water Desalting," *Proceedings of the International Symposium on Fresh Water from the Sea* (Alghero), Vol. 4, 1976, p. 301, for sale by A. A. Delyannis and A. E. Delyannis, Tsaldari St. 34, Athens-Amaroussion, Greece.

10

SUMMARY AND CONCLUSIONS

COMPARISON OF DESALTING METHODS

The planner interested in increasing the water supply to an area has to decide if salt-water purification is at all more economical than the conveyance of fresh water from the nearest available source. The decision depends on a comparison between the cost of treatment and the cost of conveyance.

Because of substantial variations in the cost of fuel, of raw materials, and of labor, it is hard to present general rules for the cost of desalting equipment and operations. Moreover, governmental subsidies for water works are common, so that comparison of the cost of desalted water and natural water is not always easy.

In areas where only sea water is available, the choice of the preferred method depends on the availability and the local cost of fuel and power. If fuel is relatively inexpensive, or if excess fuel and/or process steam are available, distillation is usually the most economical method. In the early 1970s a plant capable of producing 15,000 tons of desalted water per day cost about $10 million. In other words, the cost of installing a daily capacity of 1 ton of desalted water was about $650 for a plant of this size.

The cost of the large Hong-Kong multistage flash-distillation plant (capacity 190,000 tons/day* = 50 million U.S. gal/day) being built in 1976 was estimated at $92 million U.S., which represents an investment of *$486 per daily ton* (i.e., for each ton of the daily production capacity) or *$1.87 per daily U.S. gallon*. Cost estimates for *new* distillation plants are at least 60 percent higher. The total cost of desalted water was estimated at $2.67/kgal (= 70.5¢/ton = $869/acre-ft), whereas for *new* distillation plants in the U.S.A. realistic estimates are at least $3/kgal. In smaller plants (production capacity 3785–9500 tons/day = 1–2.5 mgd†) the total costs are higher.

It is of interest to compare these *distillation* cost figures with 1976 estimates for desalination by *reverse osmosis (hyperfiltration)*, although the product-water quality is not quite the same, because distillation yields practically salt-free water, while the hyperfiltrate usually contains several hundred ppm (parts per million) of salt.‡ For a sea-water treatment plant of daily capacity 3785 m^3 (1 mgd), the investment cost was estimated at $1,780,000 ($470 per daily ton = *$1.78 per daily gallon*) and the costs (including capital charges) at 47.5¢/ton ($1.80/kgal). (This estimate was based on interest and tax rates of 8% and 1% per year, respectively, electrical energy at 2¢/kwhr, membrane life 3 years, brine pressure 55 atm, and recovery of mechanical energy from the waste brine.) While these costs for distillation and reverse osmosis are not directly comparable because the plant size is different, other bases of comparison are not quite the same, and many factors of local significance and international commercial competition (which varies with the geographical location of the plants) were of major importance here, the figures should give the reader an idea about the capital required and the operating costs of sea-water desalting plants in 1976. The figures also show that in 1976 the costs of sea-water distillation and reverse osmosis were comparable, while distillation had been the method of choice for sea-water treatment before.

The cost of *brackish-water* desalination is in general less than of

---

*All ton designations refer to *metric* tons pure water (1 m^3 pure water weighs 1 ton and the specific weight of slightly saline water is not much different); 1 kgal = 1000 U.S. gallons.
†1 mgd = 1 million U.S. gallons per day = 3785 tons/day.
‡Hence desalting costs are sometimes computed in terms of tons of salt removed from the raw water rather than as tons of fresh water produced.

SELECTED LITERATURE

English, J. M., and El-Ramly, N. A., "Economic Evaluation of Desalting Subsystem as a Part of the Total Water System," *Desalination* **3**:308 (1967).

Koenig, L., "Economic Boundaries of Saline-Water Conversion," *J. Am. Water Works Assoc.* **51**:845 (1959).

Skrinde, R. T., and Tang, T. L., "Operating Results of Electrodialysis and Reverse-Osmosis Municipal Desalting Plants," *J. Nat. Water Supply Improvement Assoc.* **1**:15 (1974).

Channabasappa, K. C., and Strobel, J. J., "Status of Sea-Water Reverse-Osmosis Membrane Process Technology," *Proceedings of the Fifth International Symposium on Fresh Water from the Sea* (Alghero, Sardinia), Vol. 4, 1976, p. 267, for sale by A. A. and A. E. Delyannis, Tsaldari St. 34, Athens-Amaroussion, Greece. This article contains tables of operating costs for both distillation and sea-water reverse osmosis, the latter with breakdown of costs into different categories.

Roberts, E. B., and Hagan, R. M., *Energy Requirements of Alternatives in Water Supply, Use and Conservation: A Preliminary Report,* Contribution No. 155, California Water Resources Center, University of California, Davis, 1975.

Taylor, I. G., and Haugseth, L. A., "Yuma Desalting Plant Design," *Proceedings of the First Desalination Congress of the American Continent* (Mexico City), Vol. 1, Elsevier, Amsterdam, 1976, p. V-4.

Drake, F. A., "Desalting in Hong Kong—The First Phase," *Desalination* **18**:1 (1976).

APPENDICES

APPENDIX 1A: THE MINIMUM POWER FOR SPLITTING SEA WATER INTO FRESH WATER AND BRINE

The derivation of the minimum free energy for sea-water desalting in Chapter 3 is simplified and leads to the approximate value with a minimum of computation. A more detailed and general treatment is presented here.

The energy required to split salt water into fresh water and brine, all eventually at the same temperature, is *available energy*. Often it is not apparent at first sight that this energy requirement exists. Thus distillation does not require the use of mechanical or electrical energy. In this process, however, heat is transferred from the high temperature of the boiler to the low temperature of the condenser. This heat flow could conceivably be used to produce mechanical work instead and is therefore equivalent to an expenditure of *available* energy *(exergy)*.

Consider two compartments containing salt water and fresh water, respectivly, separated by a membrane permeable to water but not to salt (Fig. A.1). Almost ideal membranes of this kind are known (Chapter 9). The temperature is held constant. In equilibrium, a pressure must be exerted on the solution to prevent diffusion of water through the membrane into the solution, and resulting dilution. This is the osmotic pressure, P_{os}.

It is useful to express the osmotic pressure in terms of the activity of the water in the solution. According to elementary texts of chemical

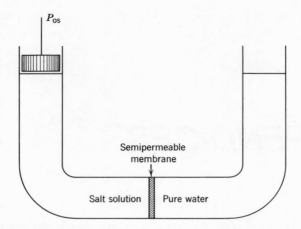

Fig. A.1. Osmotic "equilibrium" across a semipermeable membrane (a membrane permeable to water but not to salt).

thermodynamics, the chemical potential, μ, of water in solution is given by the general expression:

$$\mu = \mu_0 + P\bar{v} + RT \ln a \tag{1}$$

where μ_0 is the chemical potential of pure water at 1 atm pressure, P the pressure, in excess of 1 atm, acting on the solution, $\bar{v}$ the molar volume of water, R the gas constant, T the absolute temperature, and a the activity of the water, which decreases with increasing salt concentration. At the membrane interface, the water on either side is in equilibrium. Hence the chemical potential of water on the two sides is the same:

$$\mu_0 + P_{os}\bar{v} + RT \ln a = \mu_0 \tag{2}$$

$$P_{os} = -\frac{RT}{\bar{v}} \ln a \tag{3}$$

To separate fresh water from the solution, the pressure on the salt solution is now increased. This is to be done under reverrsible conditions. Therefore, the applied pressure exceeds the osmotic pressure by only an infinitesimal amount. The volume dv of pure water is thus pushed through the membrane, but because of the resulting increase of concentration and osmotic pressure of the solution it is soon necessary

APPENDICES

to increase the applied pressure again. Thus both the concentration of the solution and the applied pressure increase gradually until the desired amount of fresh water has been pushed through the membrane.

For each infinitesimal compression step, the work term dW equals

$$dW = P_{os}\, dv \tag{4}$$

If the initial volume of the salt solution is v_1 and the final volume v_2 liters, the total work* per liter of fresh water produced is

$$W = \frac{1}{v_1 - v_2} \int_{v_1}^{v_2} P_{os}\, dv = \frac{-RT}{(v_1 - v_2)\bar{v}} \int_{v_1}^{v_2} \ln a\, dv \tag{5}$$

To perform the integration, it is necessary to express a in terms of v. The activity is defined as

$$a \equiv \frac{p}{p_0} \tag{6}$$

where p_0 and p are the equilibrium water vapor pressures above pure water and the solution, respectively. According to many experiments (H. Sverdrup *et al.*, *The Oceans,* Prentice-Hall, New York, 1946), the vapor pressure of sea water *decreases* with increasing salinity S as shown by

$$p = p_0(1 - AS) \tag{7}$$

where $A = 0.000537$.

From equations (5), (6), and (7),

$$W = \frac{-RT}{\bar{v}(v_1 - v_2)} \int_{v_1}^{v_2} \ln(1 - AS)\, dv = \frac{ART}{\bar{v}(v_1 - v_2)} \int_{v_1}^{v_2} S\, dv \tag{8}$$

since AS is small compared to 1.

Since no salt passes through the membrane, the amount of salt Sv in the salt-water compartment (containing volume v of solution) remains

---

*Since thermodynamics evolved from man's desire to quantify and, where possible, to predict the performance of engines, work *produced* by a system was naturally counted positive, which made work *invested* negative. Equation (4) indeed takes this into account, because dv, representing the volume change of the solution, is negative. [The solution volume on the left side of Fig. A.1 continuously *decreases* upon compression ($v_2 < v_1$).]

constant from the beginning until the end of the compression process. This amount of salt is equal to the original salinity S_1 times the original weight, but as an approxomation, we shall take the (salinity × volume) product rather than the (salinity × weight) product constant:

$$Sv = S_1 v_1 \tag{9}$$

From equations (8) and (9),

$$W = \frac{-ARTS_1 v_1}{\bar{v}(v_2 - v_1)} \int_{v_1}^{v_2} \frac{dv}{v} = \left(\frac{ARTS_1}{\bar{v}}\right)\left(\frac{v_1}{v_1 - v_2} \ln \frac{v_2}{v_1}\right) \tag{10}$$

If v_2 is only slightly smaler than v_1, the logarithmic term can be converted into a simpler one:

$$\ln \frac{v_2}{v_1} = \ln\left(1 - \frac{v_1 - v_2}{v_1}\right) = \frac{v_2 - v_1}{v_1} \tag{11}$$

Hence

$$\lim W = \frac{-ARTS_1}{\bar{v}} \quad \text{for } (v_1 - v_2) \to 0 \tag{12}$$

This is the work consumed* per liter of fresh water produced when only a minor fraction of each batch of sea water is filtered through the membrane. It represents a lower limit of the energy requirements. Equation (10) shows that for higher degrees of concentration the work is larger because the expression in the second bracket is larger than unity.

For $S = 34.3‰$, $v = 0.018$ liter/mole, $T = 298°K$ (=25°C), and, by substituting the value of the constant $R = 2.31 \times 10^{-6}$ kwhr °C^{-1}mole^{-1}, we obtain

$$\lim W = -7.0 \times 10^{-4} \text{ kwhr/liter} \quad \text{or} \quad -0.7 \text{ kwhr/m}^3$$

Hence from equation (10)

minimum work (free-energy) requirement per m³ water produced

$$= -0.70 \frac{v_1}{v_1 - v_2} \ln \frac{v_1}{v_2}$$

---

*Because the work is consumed rather than produced, the work, W, as calculated from equation (12) is *negative*.

APPENDICES

TABLE A.1. Minimum Free Energy (Work) for Producing Fresh Water from Sea Water of 34.4 $^o/_{oo}$ Salinity at 25°C (kwhr/m³ fresh water)

Original volume, v_1	1	1	1	1	1
Final volume, v_2	0.99	0.7	0.5	0.33	0.25
$\dfrac{v_1}{v_s - v_2} \ln \dfrac{v_1}{v_2}$	1	1.19	1.38	1.65	1.85
Minimum free energy, kwhr m⁻³	0.70	0.84	0.97	1.16	1.30

Table A.1 lists the magnitudes of these minimum amounts of work for various degrees of water removal from the original solution, provided that each step of the salt-removal operation is carried out in ideally reversible manner.

For water of lower salinity, the minimum energy decreases in proportion to the salinity S, as can be seen from equation (10). The factor A depends somewhat on the relative proportions of the salts. The value given here refers to waters containing salts in the same relative proportions as sea water.

APPENDIX 2A: MINIMUM HEAT REQUIREMENT

The energy values in Table A.1 refer to mechanical or electrical energy, but not to heat, which is of lower value as an exergy source because only a portion of it can be converted into mechanical or electrical energy, while electrical energy can be completely converted to heat. The fraction of heat that can be converted to mechanical work in an ideal heat engine ("Carnot engine") working between the absolute temperature T and T_0 of heat source and heat sink, respectively, is*

$$\text{Carnot efficiency} = \frac{T - T_0}{T} \quad (13)$$

For instance, in an ideal steam-powered heat engine working between 130°C and 30°C (403.4° and 303.4° absolute, respectively), the

---

*Elementary thermodynamics texts, e.g., *Chemical Engineering Thermodynamics* by B. F. Dodge (McGraw-Hill, New York, 1944) and *Engineering Thermodynamics* by J. B. Jones and G. A. Hawkins (Wiley, New York, 1963), discuss this matter in detail.

Carnot efficiency is 0.248. The heat of condensation of saturated steam at 130°C is 518.5 kcal kg^{-1}. Hence each metric ton of this steam can at the most produce 518,500 × 0.248 = 128,500 work kcal = 149 kwhr. Normal steam engines do not operate on the Carnot cycle; they are not ideal and produce still less work.

Since the theoretical minimum free energy for producing 1 m^3 of fresh water from sea water at 50% recovery is 0.97 kwhr, a hypothetical ideal heat engine operating between 130°C and 30°C coupled with an ideal distillation plant should be able to produce 154 m^3 of fresh water per ton of steam. In practice, 10 m^3 of water per ton of steam is considered a good yield, for heat losses are inevitable if the process is to take place at a finite rate instead of the infinitely slow rate in a thermodynamically ideal plant.

APPENDIX 3A: WORK REQUIREMENTS FOR COMPRESSION DISTILLATION

To calculate the work done by an adiabatic compressor, we consider a piston-type compressor and algebraically add the work terms in a complete cycle. This matter is discussed in detail in the thermodynamics text by B. F. Dodge *(loc. cit.)*. Work done in any part of the cycle is counted positive here, work gained negative.

The work necessary to compress a volume of gas V_1 of pressure p_1 to volume V_2 and to pressure p_2 is

$$W_1 = \int_{V_2}^{V_1} p \, dV \qquad (V_2 < V_1) \tag{14}$$

To discharge the compressed gas at the higher pressure p_2, the work $p_2 V_2$ is necessary. When a fresh amount of gas at the lower pressure p_1 is sucked in, the work $p_1 V_1$ is gained. Since adiabatic compression raises the temperature of the gas, $p_1 V_1 \neq p_2 V_2$.

The total work done by the compressor in each cycle is therefore

$$W_c = \int_{V_2}^{V_1} p \, dV + p_2 V_2 - p_1 V_1 \tag{15}$$

APPENDICES

From the general laws of calculus,

$$d(pV) = V\,dp + p\,dV \tag{16}$$

Integrating $d(pV)$ between p_1, V_1 and p_2, V_2, we obtain

$$p_2 V_2 - p_1 V_1 = \int_{V_1}^{V_2} p\,dV + \int_{p_1}^{p_2} V\,dp \tag{17}$$

From equations (15) and (17),

$$W_c = \int_{p_1}^{p_2} V\,dp \tag{18}$$

The relation between p and v in an adiabatic compression is given by

$$pV^k = p_1 V_1^k \tag{19}$$

where k is the ratio of the specific heats of the gas at constant pressure and volume, respectively. For water vapor, k equals about 1.32.

Substituting V from equation (19) in the integral (18), we obtain

$$W_c = p_1^{1/k} V_1 \int_{p_1}^{p_2} p^{-1/k}\,dp = \frac{k}{k-1} p_1 V_1 \left[\left(\frac{p_2}{p_1}\right)^{(k-1)/k} - 1\right] \tag{20}$$

This is the exact expression for the work required by an *ideal* adiabatic compressor. For a *nonideal* compressor, W_c is divided by the compressor efficiency ϵ, which is usually about 0.75.

In compression distillation, the compression ratio $r \equiv p_2/p_1$ is close to unity, and the following simplification may be made in equation (20), with only minor loss of accuracy, by binomial expansion of the exponential

$$\left(\frac{p_2}{p_1}\right)^{(k-1)/k} = \left(1 + \frac{p_2 - p_1}{p_1}\right)^{(k-1)/k} = 1 + \frac{k-1}{k}\left(\frac{p_2 - p_1}{p_1}\right) \tag{21}$$

Hence equation (20) becomes

$$W_c = (p_2 - p_1) V_1 = p_1(r - 1) V_1 = \frac{p_1 V_1 (r^2 - 1)}{r + 1} \tag{22}$$

If (for small compression ratios), instead of $(r + 1)$ in the denominator, $2r$ is written, the range of compression values over which the simple equation approximates equation (20) is further extended:

$$W_c = p_1 V_1 \frac{r^2 - 1}{2r} = RT_1 \frac{r^2 - 1}{2r} \tag{23}$$

Equation (23) approximates equation (20) within 2% up to a compression ratio $r = 1.2$.

For example, equation (23) may be applied to twice-concentrated sea water, boiling at 101.05°C. Substituting in equation (23)

$$R = 2.31 \times 10^{-6} \text{ kwhr deg}^{-1} \text{ mole}^{-1} \qquad T = 374.4°K$$

and taking into account that 1 m³ of water contains 55.5×10^3 moles, the work for producing 1 m³ of water by this method is

$$W_c = 24.0 \times \frac{r^2 - 1}{r} \text{ kwhr/m}^3 \text{ fresh water} \tag{24}$$

The vapor pressure of the boiling brine is 1 atm at this temperature, and the compression ratio r is therefore numerically equal to the pressure (in atmospheres) prevailing in the tubes in which the vapor condenses. We can plot W_c against this pressure or against the steam condensation temperature which corresponds to it. Figure 5.12 is a plot of the latter kind for a compressor of perfect mechanical efficiency.

The line is drawn from the boiling temperature of the brine (101.05°C) to higher temperatures. It is seen that even at 101.05°C the adiabatic compression work is higher than the theoretical work for desalination which at this temperature is 1.23 kwhr/m³ (equation 10). This is because the adiabatic compression produces more heat than is necessary to raise the temperature of the vapors to 101.05°C. In practice, this excess of heat makes up for heat losses in the process, e.g., heat lost with the air vented, with the brine and distillate discharged at a higher temperature than the feed, and for losses through the insulation. From a theoretical standpoint, these effects represent not only heat losses but also a degradation of mechanical to thermal energy.

APPENDIX 4A: RELATIONSHIP BETWEEN EVAPORATION AND AIR-CIRCULATION HEAT LOSS IN A SOLAR STILL

When brine evaporates in a solar still, the vapors are transported to the cooler glass cover by convection, because of the turbulent air motion resulting from the temperature difference between brine and glass. Enhanced gas circulation in the still causes both increased water vapor transport from brine to cover, i.e., distillation, and increased heat loss by transport of sensible heat in the same direction. In fact, even if there were no convection at all and both transport phenomena were maintained only by the slow process of diffusion, they would still increase or decrease together. The relationship between the two processes is given by the ratio

$$\frac{\text{heat carried by convection}}{\text{heat carried by evaporation}} = J \frac{T_W - T_A}{P_W - P_A} \frac{P}{760} \qquad (25)$$

where T_W and T_A are the temperatures (in °C), P_W and P_A are the partial pressures of water vapor, in the saturated layers in immediate contact with brine and glass, respectively, and P is the barometric pressure (all pressures in millimeters of mercury). J is somewhat dependent on the turbulence of the air but varies only within narrow limits and can be taken as 0.46 for free evaporation and for the solar still.

The relationship between heat carried by convection and evaporation, respectively, has been known for a long time and is used in practice for the measurement of humidity by the "wet and dry bulb" method. A physicist's viewpoint on this problem can be found in the article by I. S. Bowen, "The Ratio of Heat Loss by Conduction and by Evaporation from any Water Surface" (*Phys. Rev. 27(B):*779, 1926). This ratio also plays a significant role in physical oceanography (H. U. Sverdrup, *Oceanography for Meteorologists,* Allen and Unwin, London, 1945, p. 63).

APPENDIX 5A: PROPERTIES OF ION-EXCHANGE MEMBRANES

Approximate ranges of the properties of ion-exchange membranes are listed in Table A.2. A more complete list with the names and addresses of many manufacturers may be found in the chapter "Electrodialysis" by L. H. Shaffer and M. S. Mintz in *Principles of Desalination* (Academic Press, New York, 1966, pp. 210–211).

The *transport number* of the "counterion," which is defined as the positive ion and negative ion in cation- and anion-permeable membranes, respectively, is the fraction of the current carried by that ion. The remainder is carried by the small amount of mobile ions of opposite

TABLE A.2. Some Properties of Ion-Selective Membranes for Electrodialysis

Property	Order of magnitude	Remarks
Thickness	0.1–0.7 mm	
Burst strength[a]	50–150 psi[b]	Determines resistance to sudden pressure changes
Specific gravity	1.1–1.8	
Transport number of counterion	0.90–0.99 in 0.1 N NaCl solution (5845 ppm NaCl)	Depends on concentration of solution in contact with membrane
Exchange capacity	0.5–2.0 milliequivalent per gram wet membrane	
Resistance of unit area	1–20 ohm cm^2	In solution containing 600 ppm salt
Water content	20–50%	
Water transport	0.01–0.6 ml cm^{-2} hr^{-1} at a current density of 10 milliamp cm^{-2}	
Membrane potential	About 0.055 volt per membrane	For solutions of 5000 and 500 ppm NaCl on two sides of membrane, at 25°C

[a] Measured with a "Mullen burst strength tester," American Society of Testing Materials Procedure D774.
[b] 1 psi = 0.0703 kg cm^{-2}.

charge in the membrane. Since a membrane should exclude as many ions as possible of the latter kind, the transport number of the counterions in a good ion-selective membrane should be close to unity. When the membrane is placed in concentrated solutions, this is often not the case, but some membranes retain their high selective permeability for either positive or negative ions even in fairly concentrated solutions, e.g., some Japanese membranes which were specially developed for work with sea water. These membranes were originally developed for concentrating salt from sea water rather than producing fresh water. Compare this situation with the problems of the Dead Sea industry in Israel where salt is essentially a waste product and fresh water at a premium!

The *exchange capacity* measures the concentration of counterions. Other things being equal, the higher the exchange capacity, the more selective the membrane. In a practical unit the *resistance of unit area* determines the ohmic losses in the membrane.

The *water-transport* data refer to water carried along by the ions. This phenomenon is often referred to as *electroosmosis*.

The *membrane potential* is the potential difference measured between two calomel electrodes placed in solutions of different concentration separated by the membrane. It depends on the degree of selective permeability of the membranes and should be close to its theoretical maximum value of 0.059 volt at 25°C for the solutions specified in the table. Laboratory measurements of membrane potentials are often used to screen new membrane products for selective permeability because these measurements are rapid and simple. However, for prediction of membrane performance in plants it is best to test the counterion transport number in a laboratory cell *under current* and not to rely on membrane potential data alone.

APPENDIX 6A: USE OF HEAT TO PRODUCE COLD

To calculate the amount of heat of saturated steam needed to transfer 1 cal from the refrigerated space at temperature T_2 to the ambient temperature T_0, it is well to consider a combination of two ideal engines, namely, a refrigerator which consumes mechanical energy W

and a power generator cooled by water at the ambient temperature, which produces this energy (B. F. Dodge's treatise, *loc. cit.*). The word *ideal* refers to completely reversible operation and the temperatures are measured on the absolute scale. The transfer of heat to a higher temperature is often called *heat pumping*. Note that the heat delivered at the higher temperature is higher (by the amount of compressor work) than the heat released by the low-temperature reservoir.

The work needed by the refrigerator to pump 1 cal is

$$W = \frac{T_0 - T_2}{T_2} \qquad (26)$$

The work produced by the power generator which takes in Q_1 calories at the steam temperature T_1 is

$$W = Q_1 \frac{T_1 - T_0}{T_1} \qquad (27)$$

(Both W and Q_1 are expressed in calories.)

Hence the amount of steam heat Q_1 needed per calorie pumped is

$$Q_1 = \frac{T_1}{T_2}\left(\frac{T_0 - T_2}{T_1 - T_0}\right) \qquad (28)$$

For instance, for temperatures of 101°C, 18°C, and −3°C of steam, cooling water, and refrigerated space, respectively, $Q_1 = 0.35$; i.e., 1 steam-heat calorie can remove $1/0.35 = 2.85$ cal from the refrigerated space.

Any other completely reversible cooling device, e.g., a hypothetical *ideal* absorption refrigerator, would have the same cooling efficiency when operating at these temperatures.

APPENDIX 7A: MEANS OF EXPRESSING SALT CONCENTRATION IN SEA WATER

Definitions:

Salinity: Total amount of solids (in grams) per kilogram of sea water when all carbonate has been converted to oxide, bromide and iodide have been replaced by chloride, and all organic matter has been completely oxidized.

Chlorinity: The total amount of chlorine (in grams) contained in 1 kg sea water, after all the bromide and iodide have been replaced by chloride.

Chlorosity: Total amount of chlorine (in grams) contained in 1 liter sea water of temperature 20°C after all the bromide and iodide have been replaced by chloride.

These definitions of chlorinity and chlorosity are applicable only to ocean water, normal, concentrated, or diluted. They do not apply to other types of salt water.

The empirical relation between chlorinity and salinity in sea water is

$$\text{salinity} = 0.03 + 1.805 \times \text{chlorinity}$$

APPENDIX 8A: PROPERTIES OF SEA-WATER CONCENTRATES

A number of important properties of sea water concentrates are plotted in Figs. A.2–A.4. The concentrates were prepared by evaporation without precipitation of salts. These data were taken from the publication "Critical Review of Literature on Formation and Prevention of Scale," prepared by W. L. Badger *et al.* (Office of Saline Water Research and Development Report No. 25, U.S. Department of the Interior, Washington, D.C., 1959). More extensive data on many properties of sea water and its concentrates and on solutions of sodium chloride and other salts in sea water may be found in the chapter by B. Fabuss in *Principles of Desalination,* 2nd ed. (Academic Press, New York) and in the *Saline Water Conversion Engineering Data Book* (compiled for the Office of Saline Water, U.S. Department of the Interior, by the M. W. Kellogg Co., New York, 1965, for sale by Superintendent of Documents, Washington, D.C. 20402). A considerable number of more recent data can also be found in the article "Thermodynamic Properties of Sea Salt Solutions" by L. A. Bromley, D. Singh, P. Ray, S. Sridhar, and S. M. Read (*Journal of the American Institute of Chemical Engineers* ("AIChE Journal") **20**:*326, 1974.*

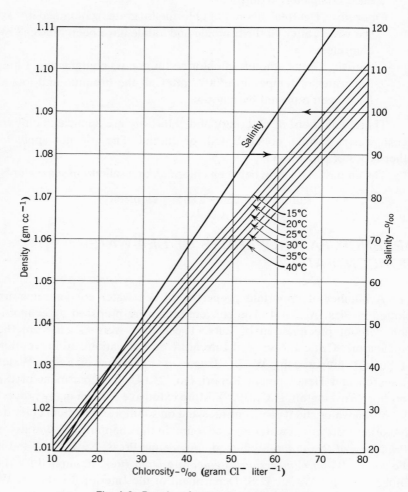

Fig. A.2. Density of sea-water concentrates.

APPENDICES

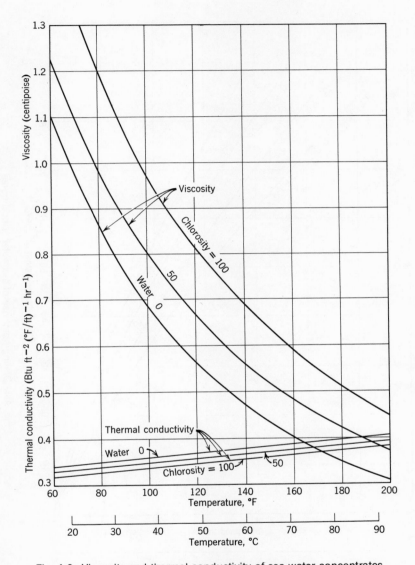

Fig. A.3. Viscosity and thermal conductivity of sea-water concentrates.

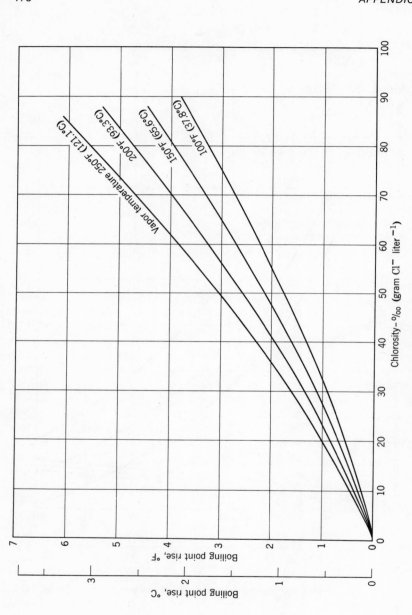

Fig. A.4. Boiling-point rise of sea-water concentrates.

APPENDIX 9A: CONVERSION FACTORS

Unit Conversion Table

To convert data in this unit ↓	to this unit ↓	multiply by factor f	$\log_{10} f$
acre (U.S. and Brit.)	square meter (m²)	4046.9	3.60713
acre-foot (acre-ft)	gallon (U.S.) (gal)	325,850	5.51302
acre-ft	cubic meter (m³)	1,233.5	3.09113
atmosphere (normal) (atm)	centimeter of mercury (cm Hg) (0°C)	76.00	1.88081
atm	inch of mercury (inch Hg) (0°C)	29.921	1.47598
atm	pound per square inch (lb/inch²)	14.696	1.16717
Btu (British Thermal Unit)	kilogram calorie (kcal)	0.25198	0.40139 − 1
Btu	kilowatt-hour (kwhr)	2.930×10^{-4}	0.46687 − 4
Btu/pound	kcal kg⁻¹	0.5556	0.74476 − 1
centimeter (cm)	inch	0.3937	0.59517 − 1
cm	foot (ft)	0.03281	0.51598 − 2
cm Hg	atm	0.01316	0.11926 − 2
cubic foot (U.S.)(ft³)	liter	28.316	1.45203
cubic meter (m³)	acre-ft	8.107×10^{-4}	0.90887 − 4
m³	gal (Imp.)	220	2.34237
m³	gal (U.S.)	264.17	2.42188
foot	cm	30.48	1.48401
gallon (U.S.) (gal)	liter	3.7853	0.57810
gal (U.S.)	gal (Imp.)	0.83268	0.92048 − 1
gal (Imp.)	gal (U.S.)	1.2009	0.07952
grain (gr)	gram (g)	0.06479	0.81158 − 2
gram (g)	grain (gr)	15.4324	1.18849
g/liter	parts per million (ppm)	1000	3.00000
horsepower (hp)	kilowatt (kw)	0.7457	0.87256 − 1
inch	cm	2.5400	0.40483
inch Hg (0°C)	atmosphere (atm)	0.033421	0.52402 − 2
kilogram (kg)	pound (avoirdupois)	2.2046	0.34333
kilogram calorie (kcal)	kwhr	0.0011628	0.06551 − 3
kcal	joules (abs.)	4,186	3.62180
kcal kg⁻¹	Btu/lb	1.80	0.25527

Unit Conversion Table continued

To convert data in this unit ↓	to this unit ↓	multiply by factor f	$\log_{10} f$
kilometer (km)	mile (statute)	0.62137	0.79335 − 1
km	yard	1,093.6	3.03886
kilowatt (kw)	horsepower (hp)	1.3410	0.12743
kilowatt-hour (kwhr)	Btu	3,413	3.53314
kwhr	joules (abs.)	3.600×10^6	6.55630
kwhr	kcal	860	2.93450
kwhr	liter-atm	35524	4.55052
liter	gal (U.S.)	0.26418	0.42190 − 1
liter-atmosphere	kwhr	2.815×10^{-5}	0.44948 − 5
logarithm to base 10 ($\log_{10}$)	ln	2.3026	0.36222
logarithm to base e (ln)	$\log_{10}$	0.4343	0.63779 − 1
meter (m)	ft	3.2808	0.51598
m	yard	1.0936	0.03886
mile (U.S. statute)	km	1.6094	0.20667
pound (avoirdupois) (lb)	kg	0.4536	0.65667 − 1
lb/inch^2	atm	0.06805	0.83280 − 2
square foot (ft^2)	m^2	0.09290	0.96802 − 2
square kilometer (km^2)	acre	247.10	2.39287
square meter (m^2)	ft^2	10.764	1.03198
ton* (metric)	ton (short)	1.1023	0.04230
ton (short)	ton (metric)	0.9072	0.95770 − 1
yard	m	0.9144	0.96114 − 1

*All references in the text are to metric tons, unless stated otherwise.

Conversion Factors for Concentration Units*

To convert concentrations given in this unit ↓	to this unit (equivalent) ↓	multiply by factor f	$\log_{10} f$
ppm calcium, Ca	ppm $CaCO_3$	2.495	0.39707
ppm magnesium, Mg	ppm $CaCO_3$	4.112	0.61404
ppm sodium, Na	ppm $CaCO_3$	2.174	0.33730
ppm potassium, K	ppm $CaCO_3$	1.279	0.10679
ppm chlorine, Cl	ppm $CaCO_3$	1.410	0.14935
ppm sulfate, SO_4	ppm $CaCO_3$	1.042	0.01770
ppm bicarbonate, HCO_3	ppm $CaCO_3$	0.819	0.91350 − 1
ppm sodium chloride, NaCl	ppm $CaCO_3$	0.855	0.93219 − 1
ppm calcium carbonate, $CaCO_3$	ppm NaCl	1.169	0.06781

*The reduction of the concentrations of all ions or salts in solution to $CaCO_3$ is done in order to find a common denominator for all of them. "50.00 ppm $CaCO_3$ simply means 1 milliequivalent per liter. Thus 12.2 mg Mg, 20.0 mg Ca, and 58.5 mg NaCl are all expressed as equivalent to 50.0 mg $CaCO_3$, because they are all 1 milliequivalent of the respective species. Note that this is merely a convenient concentration scale not to be interpreted too literally, because $CaCO_3$ itself is practically insoluble in chemically pure water. Note also that is customary to take the equivalent weight of $CaCO_3$ as the round figure 50.00, whereas the correct figure is 50.04. The factors listed here are 50.00 divided by the equivalent weights of the respective ions. For further reference on ionic equilibria in natural waters, see *Aquatic Chemistry* by W. Stumm and J. J. Morgan (Interscience, New York, 1970).

Temperature Conversion

To convert degrees Centigrade (°C) to degrees Fahrenheit (°F), multiply by 1.8, then add 32.

To convert degrees Fahrenheit (°F) to degrees Centigrade (°C), deduct 32, then divide by 1.8.

Degrees Centigrade, absolute (also called degrees Kelvin) = °C + 273.2.

Degrees Fahrenheit, absolute = °F + 460.

INDEX

Absorption refrigeration, 123
Abstracts, 11
Acid—base generation, 115
Acid injection, 43, 80
Activated carbon, 137
Activity of water, 163
Adiabatic compression, 75, 168
Admiralty metal, 63
Aeration, 132
Agriculture, use of saline water in, 13
Air-circulation heat loss, in a solar still, 171
Ammonia, for regeneration of anion-exchangers, 136
Anhydrite, 34
Anion, definition of, 102
Anion-exchange membrane, 104
Anion-exchangers, 132
Anion permeability, 102
Antiwetting agents, 56
Applied electrochemistry, text, 117
Aquaport units, 79
Aqueducts, 159
Arava valley, 8
Aruba, distillation units, 83, 159
Auxiliary refrigerator, 122, 124
Availability, 27; *see also* Utilizable energy, Exergy
Available energy, 21, 74, 105; *see also* Exergy, Utilizable energy

Back-pressure turbine-generator, 82
Baltic Sea, 19

Barium, scale, 114
Barley, salt-resistant, 8, 13
Basket-type heating elements, 44
Battery, electrodialytic, 105; *see also* Electrodialysis, reverse
Bicarbonate, 36, 38, 132
Binary fluid cycle, 84
Biomembranes, 104
Bisulfite, 147
Boilers, high-pressure, 131
Boiling-point elevation, 51, 62, 178
Boiling temperature, 48, 50
Brackish (salty) water, 2, 18
Brackish-water desalination, cost of, 156
Break-even cost for conversion, 159, 160
Brine—discharge screens, 128
Brine, disposal of, 159
Brine temperature, variation of, 90
Burst-strength, membranes, 172
Butane, 124, 130

Calcium carbonate, 16, 29, 37, 39, 63
  scale, 35
  solubility of, 37
Calcium sulfate, 16, 29
  scale, 33
  scale, in electrodialysis, 114
  solubility of, 30
Calomel electrodes, 173
Capacity, ion-exchange resins, 134, 135
Carbon, activated, 137
Carnot cycle, 167, 168

INDEX

Caspian Sea, 19
Cation, definition of, 102
Cation exchanger, 104, 131
Cellulose acetate membranes, Loeb–Sourirajan type, 143
Cellulose diacetate-triacetate blends, 147
Centers of crystallization, 31
Centrifugal compression stills, 80
Chemical potential, 164
Chile, old solar distillation installation, 84
Chlorination, 147
Chlorinity
  definition, 16, 175
  relation to salinity, 175
Chlorosity, definition of, 175
Chromotographic separations, 134
  theory, 139
Cisterns, 159
Citric acid, 43
Clay particles, 141
Coagulation, 147
Coalinga, California, electrodialysis plant, 115
Colorado River, salinity control, 2, 18, 149
Combination processes, 84
Comparative evaluation of desalting units, 116
Compression distillation, 74
  centrifugal, 81
  work requirements for, 168
Compression ratio, 169, 170
Compressor, 75
  flexible-blade, 80
  steam-ejector, 74
Concentrate, 102
Concentration, molar, 33
Condensation, dropwise, 56, 95
Condenser, 47, 59, 124
Consumption, water, domestic, 9
Contact-stabilization method, 44
Continuous ion-exchange, 134
Convection, 86
Conversion factors, 179
  for concentration units, 182
Cooper Peddy, Australia, solar distillation plant, 96
Corn starch, 44

Corrosion resistance, 63
Cost data, *see* Economics
Costs for distillation and reverse osmosis, 156
Countercurrent flow, distillation, 61
  heat transfer, 151
  ion exchange, 134, 139
Counterflow regeneration, ion exchange, 139
Counterion, 172
Countervoltage, in electrodialysis, 106
Critical point, water, 151
Critical-pressure distillation, 151
Critical velocity, in electrodialysis, 114
Crop irrigation, 158
  production of grain from, 158
  production of oranges from, 158
Crystallization, centers of, 31
Crystallizer, 120
Cylindrical mirrors, for concentration of solar radiation, 96

Dead Sea, 19, 143, 173
Deaerator, 47, 61, 67, 123, 124
Decanter, 125
Demineralization, 131; *see also* Desalination
Density, sea-water and concentrates, 176
Desalination, brackish-water, cost of, 156
Desalting plant inventories, 13
Diluate, 102
Direct-contact distillation process, 71
Direct-displacement, pump, 75
Direct-freezing processes, 120
Disodium phosphate, 44
Distillation, 47
  combination with power production, 82
  cost of, 156
  critical-pressure, 151
  flash, 64
  heat sources for, 56, 65, 69, 70
  heat-transfer coefficients, 52
  long-tube vertical, 49
  multiple-effect, 59, 60
  principle of, 48
  single-effect, 48, 58
  solar, 84
  vapor compression, 58, 74; *see also* Compression distillation
  vapor-reheat, 71, 75, 78

INDEX

Donnan effect, 143
Drinking water, maximum permissible concentration of salts, 6
Drip (trickle) irrigation, 7
Dropwise condensation, 56, 95
Dual-purpose plants (combination plants), 83
Dual water systems, 9
"Dynamic membranes," 146
Dynel, 110

Economics, break-even costs, 155, 156, 160
Effects, definition of, for distillation, 59
  optimum number of, 62
Eilat, desalination plants, 159
Ejector, 47
Electrical ion filters, 110
Electrical power, conversion of heat into, 26
Electrical power sources, 116
Electrochemistry, general, 117
Electrodes, electrodialysis, 102, 113
Electrodialysis, 101; *see also* Membranes, ion-exchange
  acid-base generation, 115
  Corfu, Greece, plant, 112
  polarization, 113
  principle, 103
  reverse, 105
  spacers, 111
  "water-splitting," *see* Electrodialysis, acid–base generation
Electroosmosis, 173
Energetics, 27
Energy, available, 21; *see also* Exergy, Utilizable energy
Equilibrium vapor pressure, definition, 22
Equivalent, chemical, 182
Evaporator, multiple-effect, 67, 68
  flash, 66
  single-effect, 57
  solar, 84
Exchange capacity, ion-exchangers, 134, 172, 173
Exergy, 21, 27, 139; *see also* Available energy, Utilizable energy
Exergy dissipation, 106
Exergy requirement, minimum, 22

Fans, rotary, 122
Ferric chloride, for scale control, 43
Fibrocement, for solar stills, 92, 96
Filtration, molecular, 146
Fixed costs, description of, 62
Flash distillation, 64
Flexible-blade compressors, 80
Fluid cycle, binary, 84
Fluted tubes, for stills, 55
Focusing devices, for solar radiation, 86
Freezing processes, 119
  auxiliary refrigeration, 121
  direct absorption, 123
  direct vapor-compression, 121
  ice-slurry washer, 128
  indirect, 125
  washer-melter, 121
Freezing tower, 120
"Freon 114," in freezing units, 124, 126
Fresh-water transportation, costs of, 158
Frothing agents, 55
Fuel needed to evaporate $1 m^3$ water, 57
Fuel values, 57

Gas constant, 164
Gas-liquid absorption, 134
Gas, noncondensing, 74
Geothermal heat, 84, 98
Greenhouse-type agriculture, combination with solar distillation, 96
Ground water, brackish, 2, 18

"Hagevap LP," 67
Hardness, 19, 30, 42, 133
Heat, conversion into electrical power, 26
Heat, source and sink, 56, 167
Heating steam, 59, 38
Heat of combustion, various fuelds, 57
Heat of condensation, 59, 168
Heat of melting, 51
Heat pump, 174
Heat-transfer coefficient, 55, 81, 133
  centrifugal-compression still, 80
  overall, 55, 67
Heat transfer of solar still compared to regular, 87
Hemihydrate, 35
Hickman No. 5 still, 81
High-pressure boilers, 131

Hollow fine-fiber units, reverse-osmosis, 146
Hong-Kong multistage flash-distillation plant, 156
Humic acids, 146
Hydrate process, 120, 129
Hyperfiltration, 141, 156; *see also* Reverse osmosis, Ultrafiltration
Hyperfiltration, sea water, 147

Impoundment, 159, 160
Indirect-freezing process, 120, 125, 129
Indirect melting, 126
Investment costs, 62
Ion-exchange, 131, 133, 139
  desalting kits, 7, 138
  membranes, properties of, 173
  membranes; *see also* Electrodialysis
  resins, 151
  resins, mixed bed, 133
  resins, regeneration, ammonium salts, 140
  resins, regeneration, evaporator brine, 139
  resins, regeneration, hot water, 138
  resins, regeneration, "Sirotherm process," 140
  resins, softening, 133
Ionic product, 33
Ions, 15
Ion-selective membrane, 104
Iron oxides, 29
Irreversibility, 24
Irrigation, of arid areas, 8
  drip (trickle), 7
Irrigation drain water for power-plant cooling, 139

Joule heat, 111

Kellogg Saline-Water Conversion Engineering Data Book, 175
Kilograins calcium carbonate per cubic foot of column space, resin capacity given in, 135
Kit, sea-water desalting, 136
Kuwait, distillation plants, 64, 97, 159

Lake Erie, 18
Latent heat of condensation, 51

Latent heat of melting, 51
Lattice, plastic, for electrodialysis spacers, 112
Lignin sulfonic acid derivatives, 44
Lime-softening, 147
Literature, salt-water purification, general, 10
Lithium chloride solution, in freezing process, 124
Long-tube vertical (LTV) evaporator, 49
Low-pressure steam, 68
Low-temperature reactors, 71

Magnesium chloride, 30
  solubility of, 30
Magnesium hydroxide, 35, 39, 63
  solubility product of, 35
Magnesium oxide, 16, 29
Membranes, *see* Electrodialysis, Reverse osmosis
Membranes, biological, 110
  compaction, 148
  dry storage, 148
  dynamic, 146
  hyperfiltration (reverse osmosis), 142
  lifetime of, 109
  potential, 172, 173
  replacement, 110
  semipermeable, 163
Membranes, ion-exchange, 104
  ion-exchange, for concentrated solutions, 173
  ion-exchange, for hyperfiltration, 143
  ion-exchange, properties of, table, 172
Membranes, ion-selective, *see* Membranes, ion-exchange
Milliequivalent, 182
Minimum heat requirement, 167
Minimum power, for desalting, 21, 22, 163
Minor constituents, natural salt waters, 15
Mirrors, for concentration of solar radiation, 86, 96
Mixed resin bed, 133
Modified cellulose acetate, 143
Molar concentration, 33
Multiple-effect evaporators, 59, 67, 68

Noncondensing gases, 74
Nonideal processes, 25

INDEX

Normal solubility, 30
"Nuclear desalination" plants, 84
Nuclei, for crystallization, 42
"Nylon"-type materials, for membranes, 146

Ocean Salinity, 15, 16, 19
  and spray, 20
Office of Saline Water (OSW), see Office of Water Research and Technology
Office of Water Research and Technology, United States Department of the Interior, 10
Oil, immiscible with water, 71
Optimization, of plant performance, 62, 63, 108
  operating versus capital costs, 63, 108
Optimum current density, 109
Osmionis, 152
Osmosis, 141
Osmotic pressure, 141, 143, 163
  of solutions, definition of, 143
Overall heat-transfer coefficient, 55, 67
Overheating, in distillation units, 51

Pans, open, for brine evaporation, 90
Partial pressure, 50, 171
Performance ratio, definition of, 68
Permeate, 144
pH changes, electrodialysis, 114
pH, definition of, 36
Plant performance, optimizing, 108
Plaster of Paris, 35
Plastic solar stills, 86
Plugging index, 147
Polarity reversal, in electrodialysis, 112, 115
Polarization phenomena, electrodialysis, 113
Polarization voltage, 106
Polyelectrolytes, as scale inhibitors, 44
Polyether/amide, membranes, for hyperfiltration, 146, 148
Polyphosphates, 64
Polysulfone, as support membrane, 148
Port-Orange, Florida, solar stills, 91
Power index, electrodialysis, 106, 107
Power requirements, for desalting, 21, 22, 105, 163

Propane hydrate, 129
Pumps, direct-displacement, 75
Pyrheliometer, 86, 93

Radiation losses, 91
Radiometer, for solar-energy research, 92
Rare-earth ions, 134
Reactors, nuclear, as heat sources for distillation, 71
Recirculation systems, 105, 116
Refrigeration, absorption, 174
  auxiliary, 122
  exergy requirements for, 173
  technology of, 119
Regeneration, ion exchange resins, 133, 135
  ion exchange resins, chemicals for, 135
Resins, ion exchange, 134
Reverse osmosis, 2, 141, 156; see also Hyperfiltration, Ultrafiltration
Reverse osmosis, costs of, 156
  membranes for, 143
  sea-water treatment by, 101
  units for, 145
Reversible process, 24, 126, 174
Rotary fans, 122
Rotocell, 129

Saline water, see Salt waters
Saline-Water Conversion Engineering Data Book, 175
Salinity, definition, 16, 174
  and chlorinity, relation between, 175
  ocean, 1, 19
  specifications, 6
  specifications, for agriculture, 7
  specifications, for human consumption, 6
  specifications, for industry, 7
Salt concentration, see Salinity
Salt-depleted layers, 104
Salt Lake City, Utah, brackish well, 18
Salt rejection, 143
Salt-water conversion, 2
Salt waters, natural, composition of, 15
"Saturation index" (Langelier), 39
Scale, 29, 41, 104
  barium sulfate, 114
  calcium carbonate, 35

Scale (*cont'd*)
   calcium sulfate, 114
   definition of, 29
   delayed deposition, 43
   distillation processes, 42
   electrodialysis processes, 113
   hyperfiltration processes, 43
   magnesium hydroxide, 35
   prevention of, 42
   soft, 38
   strontium sulfate, 114
   type, dependence on temperature, 39
Scale deposition, delay of, 43
Scraping devices, 79
Sea water and concentrates, thermodynamic properties of, 27
Sea-water desalting kit, 136
Sea-water heater, 71
Sea-water hyperfiltration, 148
Sea water, minor constituents, 15
   composition of, 16
Sea water and vapor, parallel flow of, 61
Secondary butane cycle, 126
Second law of thermodynamics, 24
Seeds, 44
Selective permeability, membranes, 104, 173
Semipermeability, membranes, 141
Servel-Electrolux, refrigeration method, 123
Shenandoah River, composition of, 18
Shipping, 158
Silver ions, 136
Single-effect evaporator, 57, 59
Single-stage flash evaporator, 66
"Sirotherm," method for ion-exchange resin regeneration, 138, 140
Slurries, in freezing processes, 128
Slush, in freezing processes, 128
Snake-cage resins, 151
Soda, 44
Sodium hexametaphosphate, 44
Softening, 42
   definition of, 3
   ion-exchange, 133
Solar distillation, combination with greenhouse-type agriculture, 96
Solar energy, *see* Solar radiation
Solar energy, general journals on, 100

Solar evaporation, 84; *see also* Solar distillation
Solar radiation, incidence of, 86
   composition of, 87
   concentration of, mirrors, 96
   monthly average, 92
Solar radiation received on horizontal surface at different places, 88
Solar still, 84
   area necessary for producing irrigation water, 87
   covered, 90
   Daytona Beach test site, 94
   deep-basin, 91
   efficiency of, 90
   energy balance for, 92
   evaporation and air circulation, relation of, 171
   plastic, 95
   principle of, 85
   simple, 85
   tilted, 89
   yields, of, 93
Solubility, 30
   behavior, inverted, 31
   behavior, normal, 30
   calcium carbonate, 37
   calcium sulfate, 30, 33
   magnesium chloride, 30
   magnesium hydroxide, 35
   product, 34, 36
   sodium chloride, 30
Solvent extraction, 149, 150
Spacer, electrodialysis cells, 104
Sparging, 55
Specific gravity, 172
Specific heat, 51
Spiral-wound unit, hyperfiltration, 144, 148
Spray, ocean, 20
Stability data, 39
Stability diagram, for calcium carbonate, 40
Stability diagram, for magnesium hydroxide, 37
Starch, as a scale inhibitor, 43
Steam, 47, 51
   low pressure, 68
Steam chest, 47
Steam ejector, 61

INDEX

Steam-ejector compressor, 74
Steam-side heat-transfer coefficients, 56
Steam strippers, 151
Stefan's law, 91
Still, solar, 84, 87
Stills, centrifugal compression, 80
Stills, plastic, solar, 86
Stills, principle of, 86
Storage batteries, 107
Strong-acid resins, 136
Strong-base anion-exchange resins, 136
Sulfates, 114
Sun, as heat source, 84; see also Solar evaporation, Solar radiation
Superintendent of Documents, U.S. Government Printing Office, 10
Supersaturation, 31, 32, 34, 40, 41
Surfactants, 55
Swelling clay, 137
Symi, Greece, solar still, 96

Temperature conversion, method, 182
Thermal conductivity, different materials, 54
  sea water and concentrates, 177
Thermal energy, 21; see also Heat
Thermal insulation, 91
Thermal regeneration, ion-exchange resins, 140
"Thermal shocking," 45
Thermocompression evaporators, 41
Thermodynamic properties of sea-salt solutions, 27
Thermodynamics, second law, 24
Thermoeconomics, 27
Time lag, in solar stills, 90
Titanium, for electrodes, 113
Transportation costs, fresh-water, 158
Transport number, 172
Triethylamine, 149
Tubular units, hyperfiltration, 144
Turbines, 148, 151
  power recovery, 148
Turbulence, air, in solar stills, 171
Turbulence promoters, 56

Ultrafiltration, 145
Ultrapure water, in manufacture of television tubes, 131

Unit conversion table, 179
United States Department of the Interior, Office of Water Research and Technology, 10
United States Government Printing Office, Superintendent of Documents, 10
Use of heat to produce cold, 173
Utilizable ("usable") energy, 21; see also Available energy, Exergy

Vacuum-flash process, 120
Vapor-compression distillation, 74; see also Compression distillation
Vaporization, 51
Vapor-pressure curves, 48
Vapor pressure, equilibrium, 22
  pure water and sea water, 50
Viscosity, sea-water and concentrates, 177
Volatility of salts, 47

Wash column, ice, 127, 128
Water, distribution systems, dual, 9
  index, definition of, 4
  index, table, 5
  in ion-exchange membranes, 172
  irrigation, amount necessary, 8
  transport, in ion-exchange membranes, 172
  use of, amount, 5, 13
  vapor pressure of, 50
"Water-splitting," see Electrodialysis, acid-base generation
Waterworks, municipal, 3
Weak-acid cation-exchange resins, 135
Weak-base anion-exchange resins, 136
Wellton-Mohawk saline irrigation return flow, 2, 18, 20, 101
Wet and dry bulb temperature, 171
Wilson cloud chamber, 76
Wiped-film technique, 56
Work requirements for compression distillation, 168

Yuma, Arizona, desalination plants, 149, 157

Zarchin direct-freezing process, 79, 120
Zone freezing, 152

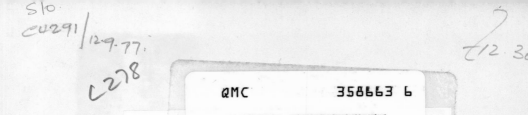
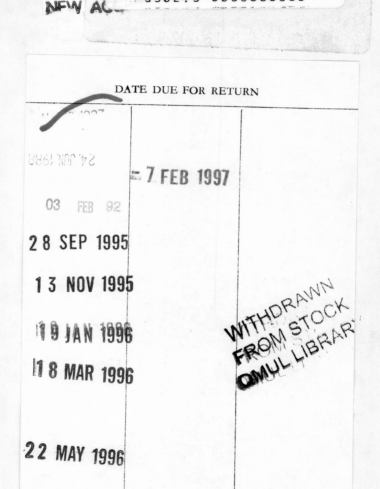